4 D 98
£14-95

CW01584465

PRINCIPLES OF ECOLOGY
in Plant Production

PRINCIPLES OF ECOLOGY *in Plant Production*

Edited by

T.R. SINCLAIR and the late F.P. GARDNER

University of Florida
Institute of Food and Agricultural Sciences
Agronomy Department
Gainesville, Florida
USA

CAB INTERNATIONAL

CAB INTERNATIONAL
Wallingford
Oxon OX10 8DE
UK
Tel: +44 (0)1491 832111
Fax: +44 (0)1491 833508
E-mail: cabi@cabi.org

CAB INTERNATIONAL
198 Madison Avenue
New York, NY 10016–4314
USA
Tel: +1 212 726 6490
Fax: +1 212 686 7993
E-mail: cabi-nao@cabi.org

A catalogue record for this book is available from the British Library, London, UK

Library of Congress Cataloging-in-Publication Data

Principles of ecology in plant production/edited by T.R. Sinclair and F.P. Gardner.
p. cm.
Includes index.
ISBN 0-85199-220-X (alk. paper)
1. Crops--Ecology. I. Sinclair, Thomas R., 1944– .
II. Gardner, Franklin P. (Franklin Pierce), 1924–1992.
SB91.P78 1998
630'.2'77--dc21

97-33999
CIP

ISBN 0 85199 220 X

Typeset in Souvenir by Columns Design Ltd, Reading
Printed and bound in the UK at the University Press, Cambridge

Contents

Contributors vii
Preface ix

Introduction

1. Human Population, Plant Production and Environmental Issues 1
K.L. Buhr and T.R. Sinclair

Plant Production

2. Ecological Perspective in Plant Production 19
F.P. Gardner and T.R. Sinclair

3. Diversity and Genetics in the Development of Crop Species 31
D.A. Knauft and F.P. Gardner

4. Development of Agricultural Ecosystems 49
D.E. McCloud

5. Environmental Limits to Plant Production 63
T.R. Sinclair and F.P. Gardner

6. Soil and Plant Production 79
E.A. Hanlon and F.M. Rhoads

7. Water 103
T.R. Sinclair and J.M. Bennett

Plant Environment

8. Radiant Energy 121
T.R. Sinclair and F.P. Gardner

9. Temperature 135
K.J. Boote and F.P. Gardner

10. Weather and Climate 155
L.H. Allen and F.P. Gardner

11. Carbon Dioxide and Other Atmospheric Gases 169
L.H. Allen

Appendix of Scientific Names 185

Index 187

Contributors

Dr L.H. Allen, Jr
Agronomy Department, Agronomy Physiology Laboratory, IFAS Building #350, University of Florida, PO Box 110965, Gainesville, FL 32611–0965, USA

Dr J.M. Bennett
Agronomy Department, 304 Newell Hall, University of Florida, PO Box 110500, Gainesville, FL 32611–0500, USA

Dr K.J. Boote
Agronomy Department, 304 Newell Hall, University of Florida, PO Box 110500, Gainesville, FL 32611–0500, USA

Dr K.L. Buhr
Agronomy Department, 2183 McCarty Hall, University of Florida, PO Box 110300, Gainesville, FL 32611–0300, USA

Dr E.A. Hanlon
Southwest Florida REC, PO Box 5127, Immokalee, FL 34143–35002, USA

Dr D.A. Knauft
North Carolina State University, Department of Crop Science, Box 7620, Raleigh, NC 27695–7620, USA

Dr D.E. McCloud
1660 NW 22nd Circle, Gainesville, FL 32605, USA

Dr F.M. Rhoads
University of Florida, North Florida Research and Education Center, Route 3, Box 4370, Quincy, FL 32351–9529, USA

Dr T.R. Sinclair
USDA, ARS, SAA, Agronomy Department, Agronomy Physiology Laboratory, IFAS Building #350, University of Florida, PO Box 110965, Gainesville, FL 32611–0965, USA

Preface

The production of food, fiber, and fuel has been the major challenge to humanity throughout history, and this challenge continues today. Plant production is the only basis for providing food; the diet of many people is based almost solely on direct consumption of plant products. While others can add meat to their diet, meat production is based on feed obtained through plant production. Fiber needs are also dependent on plant production either directly in the case of wood and cotton, or indirectly in the production of animal fibers such as wool. Plant products remain the main fuel for cooking and heating for many people in less-developed countries. Clearly, sustained plant production is absolutely essential in assuring the well-being of the world's population.

Furthermore, the challenge to plant production is compounded by the dramatic increases in the last half century in the world's population. Human population is increasing at an astounding rate – approximately 250,000 people per day or 90 million per year! Each new person adds to the demand for food, fiber, and fuel. Further, increases in economic well-being of any segment of the population results in increasing per capita demand for plant products.

The demands for increasing production of plant products come at a time when there are worldwide concerns about environmental issues in modern management of crop lands, grasslands, and forests. The objective of this book is to offer a textbook for students who have a professional or personal interest in environmental issues associated with sustaining, or even increasing, plant production. The intent is to provide to students, from a broad range of backgrounds and interests, basic information on the processes that define the ecology and environment of plant production systems. Many of the examples discussed in the book are drawn from field crop production, but the issues are common to all types of plant production including horticulture, forestry, and rangelands. A scientific approach is taken in discussing the various issues, although students need only a general biology background as the starting point to comprehend this book.

Dr Franklin Gardner, a renowned crop ecologist, recognized the critical need for a textbook that presents a scientific background to the many environmental issues that are in the daily headlines concerning plant production. Upon his retirement from teaching and research at the University of Florida, Dr Gardner began to assemble his career experiences into a book on the environmental and ecological issues in plant production. Unfortunately, his death left the book only half completed. Recognizing the importance of the book, several of Dr Gardner's colleagues took on the task of completing the book. The experiences and knowledge of Dr Gardner have now been augmented by specialists in each of the topics presented in this book.

The book begins with an introductory chapter that discusses issues associated with increasing human population, increasing demand for plant products, and sustaining the environment for plant production. The bulk of the book is divided into two major sections. The first section considers plant production and its relation to the environment, and the second section examines aspects of the physical environment that impact plant production systems.

The Plant Production section begins with a discussion of the general concepts and unique features of the ecology of plant production systems (Chapter 2). This chapter is followed by an examination of the domestication of plant species and the development of crop varieties (Chapter 3). Modern genetic manipulations of plants including the use of biotechnology techniques are discussed. Chapter 4 considers the historical development of various agricultural ecosystems and the factors associated with increasing crop yields. This is followed (Chapter 5) by an evaluation of the biophysical limits to plant productivity and future prospects for increasing crop yields. The final two chapters consider two environmental factors that have major influences on plant productivity. Chapter 6 discusses the importance of soils in plant growth. Topics that are considered include soil texture, soil erosion, and soil fertility. Chapter 7 evaluates the critical requirement for water in plant production systems.

The Plant Environment section examines several aspects of the physical environment of plant production systems and discusses the relationship with specific environmental issues. This section opens with a discussion of solar radiation, the energy source for life (Chapter 8). Questions about the quantity and quality of light and the impact on plant growth are considered. The subsequent chapter (Chapter 9) discusses the major influence of temperature on plant development. The potential impact of global warming on plant production is considered. Weather patterns are critical determinants in selecting plant species for production and in limiting plant growth (Chapter 10). The development of these weather patterns and their impact on plants are discussed. Finally, in Chapter 11 the effects of increasing concentrations of atmospheric carbon dioxide and gaseous pollutants on plant communities and plant growth are examined.

Human Population, Plant Production and Environmental Issues

1

K.L. BUHR AND T.R. SINCLAIR

Changes in human population are closely associated with improvements in plant production. About 10,000 years ago when food was obtained by 'hunting and gathering' the human population on Earth was about 5 million people. With the initial domestications of plants and animals, populations only increased by a small amount. Roughly 5000 years ago, with the establishment of cities, the population began to increase substantially. The milestone of 100 million people on Earth was reached about 2500 years ago. Populations continued to increase with improvements in crop production, and the expansion of cropped land area.

The threshold of modern increases in human population was reached in about 1830 when the human population reached 1 billion people. Since then, with scientific advances associated with increasing plant production, population increases have been very rapid. The world population doubled to 2 billion by about 1930 and doubled again to 4 billion by about 1975. The population is now at about 5.9 billion people with an increase of about 90 million each year.

These statistics on the size of the human population lead to similar staggering facts about the recent increases in resources needed to sustain these people. Such basic needs as food to eat, fiber for clothing and shelter, and plant-derived products for fuel are being produced at unprecedented levels. Plant production is ultimately the basis for nearly all the food, most of the fiber, and much of the fuel for many people. Consequently, sustaining and increasing plant production has been, and will continue to be, an essential concern. An important aspect of this book is the background of the development of agriculture and increasing plant production within the context of Earth's varied environments (Chapters 2 to 4).

The pressure for plant production resulting from a huge human population results in tremendous pressures on the environment. Many of today's critical environmental issues are either a result of the need to increase plant productivity or the sensitivity of plant production systems to changing environmental conditions. This book focuses specifically on some of these critical environmental issues that will influence political and economic decisions in the coming decades. Before examining each of the environmental issues associated with

CAB INTERNATIONAL 1998. *Principles of Ecology in Plant Production*
(eds T.R. Sinclair and F.P. Gardner)

plant production, the magnitudes of the demands and consequences of these demands on plant production are discussed in this chapter. What are the projected future changes in the human population? What is the interaction of food supply and increasing population? How is increasing demand for plant production associated with many critical environmental issues?

Population Trends

An important perspective in understanding population trends is to examine doubling time, that is, the length of time over which the population doubled. Two thousand years ago, there were about 250 million people on the Earth. It took about 1650 years for the population to double to 500 million. The next doubling took less than 200 years so that by 1830 Earth's human population had passed 1 billion. After that, the doubling time continued to shrink with just another 100 years required to reach 2 billion, then only 45 years more to get to 4 billion. Never before the twentieth century had any human lived through a doubling of Earth's population. Now, the Earth's human population is 5.9 billion. Various projections for the future raise the prospect of further substantial increases in populations (Fig. 1.1).

In 1965 the global population growth rate peaked at 2.1% per year (a rate sufficient to double the global population in 36 years). Since then, the growth rate has fallen to 1.5% per year, a rate that would still double the population in 48 years. For the first time in human history, the population growth rate has slowed, despite a continuing drop in death rates, because people are having fewer children. China is an important example of a country that has achieved a slowing of their population increase.

Nevertheless, the human population continues its relentless increase. In absolute numbers, the net annual increase in the world population is 88 million. This is a daily increase of about 240,000 people, or the equivalent of a moderate size city. That is, each day, worldwide plant production must be increased to provide the food, fiber, and fuel for a new city.

To make matters worse, the population increase is not uniform and there are nations where population growth rates are still at or near 4% a year. This growth rate doubles the population in just 18 years. Most of the inhabitants in these high growth rate populations have not yet even reached child-bearing age. There are countries in Africa, in particular, where there are very high population increases. As a comparison, between 1990 and 2050 China's population could rise from 1.2 billion to nearly 2 billion. Over the same period Africa's population may grow to perhaps 2.3 billion from 600 million, a 280% increase, if current fertility rates are sustained. Barring catastrophic death rates, by 2075 Africa's population could surpass the population of Asia.

More than 90% of future population growth is expected to be in countries identified as 'developing'. Most women living in these countries today are descended from generations of women who were essentially continuously pregnant throughout early adulthood. In many of these societies, bearing or

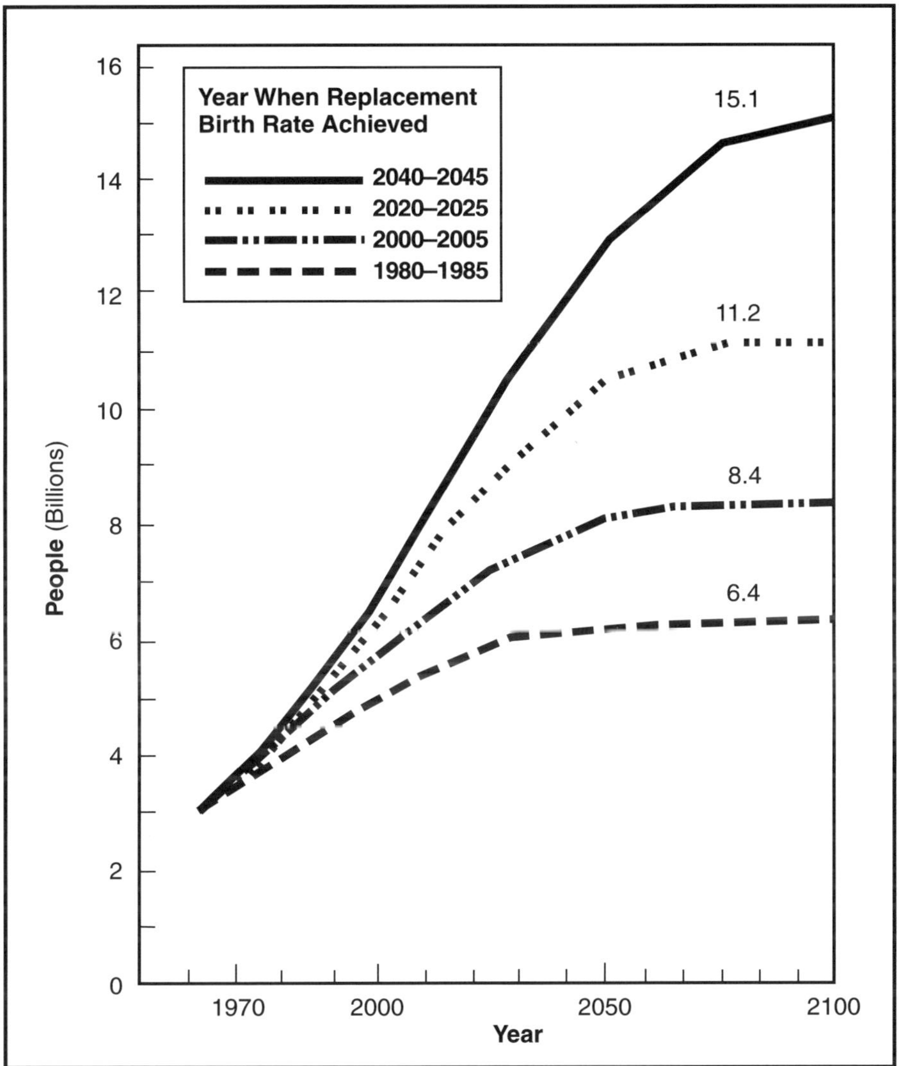

Figure 1.1 ***Projected increases in global population plotted against year based on various time periods assumed for a decrease in the human birth rate to the replacement rate.***

attempting to bear large families is important to improve the chances of one child or more surviving to provide support for the parents in their old age. At this time, the average number of births per female, worldwide, is 3.1. For the global population to stabilize, that figure should be in the 2.0 to 2.1 range. If the lower figure could be reached by the year 2000, the global population would still eventually grow to about 8.4 billion before the population stabilized (Fig. 1.1).

Limiting population growth, however, is an especially sensitive matter. Who is to say whose reproductive rights must be restricted? There are ethical and moral dilemmas, especially when rich nations attempt to influence such sensitive issues in the poorer nations. Factors necessary for families in developing countries to voluntarily reduce the number of births in their families are: improved health care, improved old age security, and more education. The education of girls and women has been especially important in limiting population growth. When they learn to read and have control over their own lives and destiny, fertility rates decline.

Food Requirements

Thomas Malthus published in 1798 *An Essay on the Principle of Population*, which warned that population growth would eventually exceed the capabilities to produce food. Malthus believed that population growth eventually would lead to poverty. Antoni van Leeuwenhoek, inventor of the microscope, made a guess at the carrying capacity of the planet. In 1679 he calculated that if all the Earth's habitable land could be settled as densely as Holland at that time, then the Earth could accommodate 13.4 billion people. While that figure may have seemed astronomical to Leeuwenhoek, it is now exceeded by one of the population increase scenarios in Fig. 1.1.

So far, farmers and science have stayed ahead of population growth. Substantial increases in rice and wheat yields resulting from the 'Green Revolution' during the 1960s and 1970s allowed more food to be made available to more people at lower costs. Per capita availability of protein and calories increased, especially in developing countries, although problems of food distribution remained. Life expectancy continues to improve in most parts of the world and infant mortality rates have declined substantially. Yet, can we be certain that Malthus was wrong, other than in his timing?

Certain areas of the world are not expected to keep pace in their supply of food production to meet rising demand. Africa, notably, is expected to need 250 million metric tonnes of grain by 2030 – 10 times current import levels. The Indian subcontinent (the site of remarkable gains from Green Revolution technologies in wheat) is expected to return to food-deficit status. Other countries with rapid population growth – among them Iran, Egypt, Ethiopia, Mexico and Nigeria – may be facing substantial food deficits in the years ahead. In China, the largest rice producer, yields (production per hectare) grew by 4% annually in the 1970s. In the 1980s, the increase in yields was 1.6%, and it is even lower in the 1990s.

After nearly four decades of expansion in food supplies, productivity gains are experiencing a loss of momentum. Between 1950 and 1984, world grain production expanded 2.6-fold (Fig. 1.2A), keeping ahead of population growth and raising the amount of grain harvested per person (Fig. 1.2B). In recent years, however, these trends have not continued. After expanding at 3% a year from 1950 to 1984, the growth in grain production per unit land area has

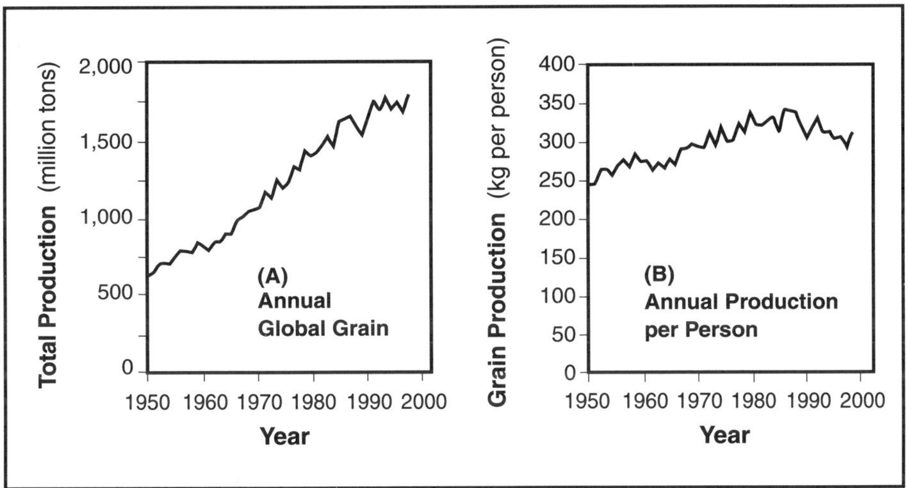

Figure 1.2 ***(A) Annual global total grain production and (B) annual global production per person plotted against year.***

slowed, rising at about 1% annually from 1984 until 1993. As a result, grain production per person fell 12% during this period (Fig. 1.2B).

Not only does increasing population increase the demand for food supply, but increasing prosperity also results in increasing food demands as dietary standards change. More animal products are eaten. During the mid-1990s about half of the increasing demand for food comes from population growth and the other half from increasing prosperity. One-third of the grain supply worldwide is devoted to livestock, two-thirds of the grain in the US goes into producing meat. It takes about seven units of grain to produce one unit of beef, four units of grain for one unit of pork, and two units of grain for one unit of chicken. In addition, grazing animals can put severe strains on the environment. For example, increased numbers of grazing animals have contributed to the degradation of grasslands south of the Sahara Desert in the Sahel region of Africa.

Fiber and Fuel Requirements

The increase in human population and improved standard of living also translates directly into increased demand for fiber and fuel derived from plant production. One major need for fiber is improved quality and quantity of clothing for more people. In addition, population pressures result in increased demand for wood products such as newsprint, lumber, and fuelwood.

Not surprisingly, the demand for plant-derived textiles, particularly cotton, has increased steadily with population. In 1950, world cotton consumption (and production) was about 7.5 million metric tonnes per year. Since then, in a period of only 45 years, the world consumption has nearly tripled! This remarkable increase in cotton production has been met exclusively by increased yield per land area, with no increase in the amount of land area sown to cotton. This achievement is even more remarkable because the pressure to increase the production of food crops has tended to push cotton production to less productive lands.

Cotton management practices have intensified in recent decades, along with potential environmental hazards, to meet the demand for cotton. Now, irrigation, fertilization, and pesticides are all common in cotton production. In the future, the pressure of increased cotton consumption will continue to challenge the production systems for higher yields while minimizing adverse environmental consequences.

Forests and woodlands cover roughly 5.5 billion hectares, or approximately one-third of the Earth's land surface. Consequently, this huge area can have tremendous influence on global environmental issues. For example, concerns about atmospheric increases of CO_2 and climate change must account for deforestation and the role of the forest in the environment. The management of forests to increase productivity also involves many environmental issues common to other plant production systems.

In developed countries many forests are regrowth forests managed for fiber production to be used as pulpwood and lumber. This is particularly true for softwood lumber and pulpwood production. The steady increase in net consumption of wood products by industrialized countries has been met mostly by increasing productivity of the forests within these countries. The industrialized countries, however, are net importers of hardwood from developing countries in the tropics.

On the other hand, increases in population and desires for improved standards of living in developing countries have resulted in greatly increased demands for all types of forest products. Some of the most spectacular increases in consumption have occurred in newsprint as the greater numbers of literate people in the developing countries increase the demand for paper. This increased demand for forest products has outpaced the production capabilities in many of these countries although they may have abundant native forests. The development of processing facilities for forest products has not been able to keep up with the increases in consumption. As a whole, less developed countries are net importers of newsprint, pulpwood, and softwood lumber.

Forests in developing countries are also particularly important as fuelwood sources for use in cooking and heating. In Africa, where forest products provide much subsistence energy supplies, forests are vanishing at a rate of 1.7% a year, twice the average for developing countries. Harvesting of fuelwood in these regions results in a common sequence of environmental degradation. It often begins when the fuelwood demands of a growing population exceed the sustainable yield of local forests, leading to deforestation. As firewood becomes scarce, cow dung and crop residues are burned for fuel, depriving the land of nutrients and organic matter. Livestock numbers expand more or less with the human population, eventually exceeding grazing capacity of the land. The combination of deforestation and overgrazing increases rainfall runoff and soil erosion, simultaneously reducing aquifer recharge and soil fertility, leading to further degradation.

Increasing fuelwood production is a high priority for forestry in many developing countries. This is particularly true in tropical regions where other energy sources have usually not been developed. Agroforestry, which uses relatively intensive management practices to encourage rapid tree growth, is gaining worldwide interest as a method to meet the fuel and fiber needs of many countries.

In developing countries, wood materials and other plant products meet only a small fraction of the energy needs. New schemes are being explored to use plant products directly as alternate energy sources to fossil fuels in these countries. Both maize and sugarcane have been used as a basis for manufacturing fuel alcohol to power automobiles.

General Consequences of Increasing Demand for Plant Products

Economic

Economically, food is a highly 'inelastic' commodity. That is, when shortages develop prices rise sharply as people seek to fulfill their daily needs. This can be illustrated by examining the changes in world grain prices in 1973, and again in 1996. Crop yields were depressed in 1973, particularly in the US, at a time when carryover stocks of grains were low (Fig. 1.3). Soybean prices tripled in this situation of decreased supplies. To make matters worse, the President of the USA enacted an export embargo on soybeans, which had more than tripled in price before the embargo. The price increase and subsequent embargo sent ripples through international markets. Commodity importers, especially soybean importers, sought alternatives to importing soybeans from the USA.

Farmers and growers worldwide responded to the increased prices following the 1973 growing season by increasing land area in soybean production. With increased production, commodity prices returned to levels at or slightly above prices before the jump in prices. With the decline in prices, concerns about food production and concerns about the carrying capacity of the Earth subsided. Within 3 years, carryover stocks had returned to comfortable levels and by the late 1970s and early 1980s, surplus production was the concern in the industrialized countries, rather than insufficient production. The situation was reversed again in the mid-1990s as carryover stocks decreased to the lowest levels in recent history (Fig. 1.3).

International trade patterns in maize offer an interesting case study in the economic consequences of this recent decline in production per person. In the mid-1990s, China as the world's second largest producer of maize moved from exporting to importing maize. The USA as the world's leading grain exporter had generally been able to fully supply the world markets. In 1995, however, the US crop was reduced for a variety of reasons. Within months, prices of maize, and other agricultural commodities experienced substantial price

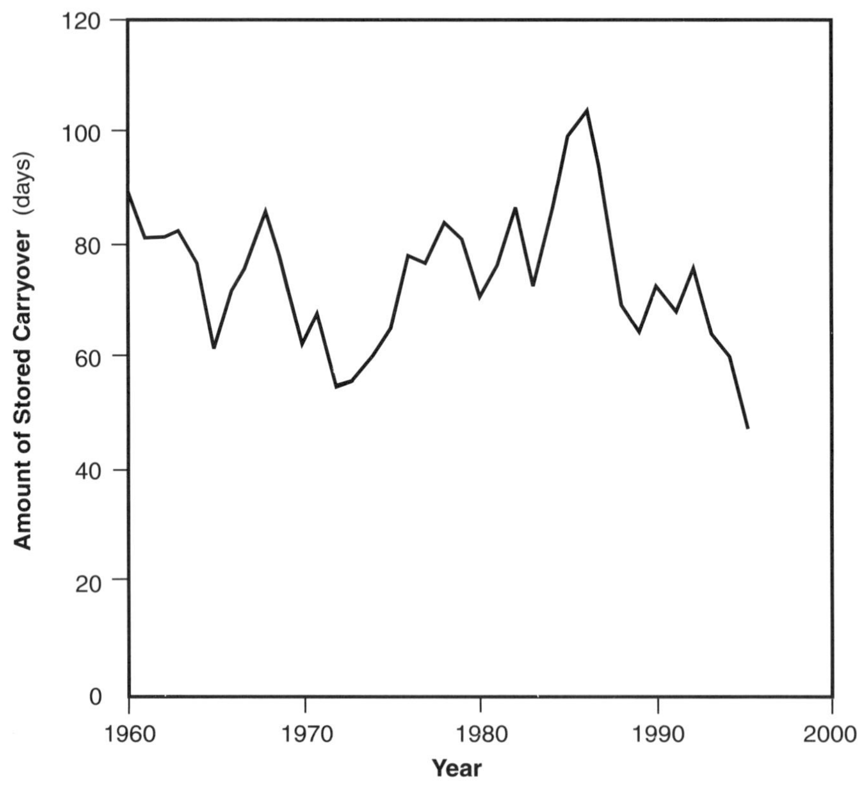

Figure 1.3 ***Amount of grain carryover stocks in the world reserves expressed in days, and plotted against year.***

increases. Prices more than doubled in maize and wheat, again raising concerns about the Earth's food production capacity.

China's 1996 grain purchases have attracted special attention because of the potential implications for future grain markets. With a population of 1.2 billion, China is adding 14 million people a year. Meanwhile, China's rapidly expanding economy is increasing personal incomes, even faster than the population, placing ever greater demand on all food supplies. While this surge in grain demand is occurring, observers question China's capacity to increase food production. Droughts, declining water tables, salinity, soil erosion and conversion of crop land to non-farm use raise concerns about China's potential to produce the food required by its growing and increasingly affluent population.

There is a demographic dilemma in those regions where the local or regional requirements for food production cannot be met within the region. This is particularly true for countries whose economies do not have the hard currency to import increasing quantities of food. The three options are: face starvation, become dependent on food aid, or emigrate. Before confronting this demographic dilemma, however, the country's initial efforts to provide itself with food, water, and fuelwood can degrade ecosystems that may take decades or centuries to recover. This has happened, or is happening, in places such as parts of Africa south of the Sahara, in the foothills of the Himalayas and the Andes, and in crowded small nations such as Haiti.

Health and well-being

The Food and Agriculture Organization (FAO) estimated that there are more than 800 million people who are malnourished, about a third of whom are children. Among malnourished children less than 5 years old, the World Health Organization estimated that 40,000 die each *day* as a direct result of starvation or as an indirect consequence of malnutrition. Of these 40,000 daily deaths, two-thirds are children who have failed to make it to their first birthday.

Africa in particular may bear the brunt of starvation and malnutrition. Not only does the continent have the highest population growth, but the increase in per capita food production has stagnated. Africa has a harsh climate, poor endowments of natural resources, inadequate public investment in research and infrastructure, political corruption and instability, and rural poverty and gender inequity. These have all conspired to keep the increase in food production at just 2% a year – below the rate of population growth. As much as 40% of the population is malnourished for at least part of each year.

Crop yields

In less than one century, agriculture has undergone a major transition from a labor-intensive to a science-based system. Scientific advances and improved grower practices have doubled grain production from 1.4 to 2.7 metric tonnes per hectare since 1945. While the global population has more than doubled, the average daily calories per person have still climbed from 2300 kcal in the early 1960s to 2700 kcal in the mid-1990s.

By doubling or even tripling the yield of several crops, US farmers have avoided the need to expand the land area under cultivation. Advances in productivity have afforded the luxury of using land as parks and nature preserves, residential lawns, golf courses, and other recreational facilities. Further, in the interests of soil conservation and erosion control, some land in industrialized countries had been taken out of crop production. In 1996, however, there was a dramatic decrease in the land area that was set aside in these land reserve programs.

Yet, in spite of the successes in increasing productivity, there are some ominous signs. The rate of crop yield increase is tapering off both globally (Fig. 1.4) and in regions where farmers have been applying modern agricultural methods. For example, during the 1960s, US wheat gained in yield over the 1950s by an

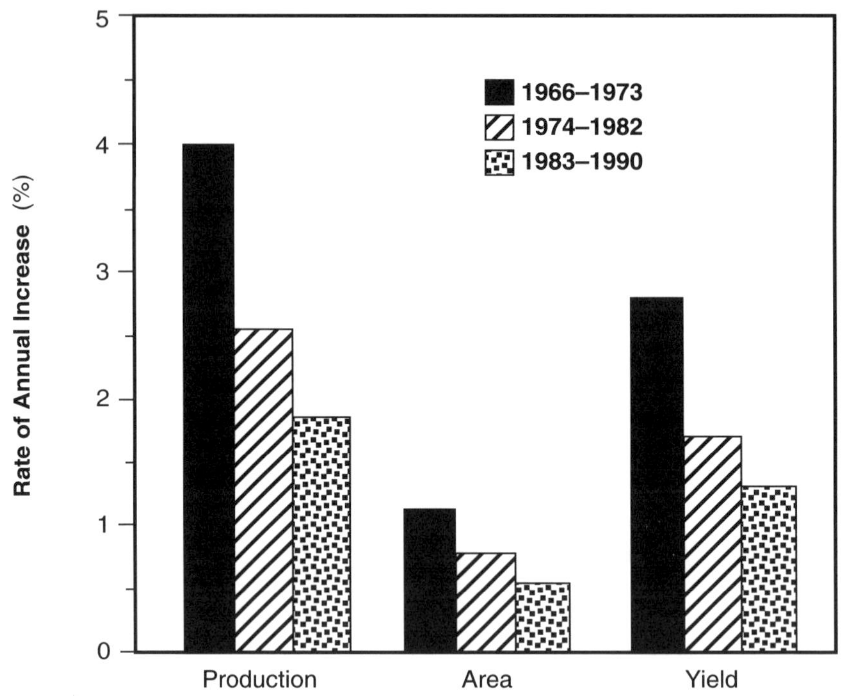

Figure 1.4 ***Rate of annual increase in crop production, crop land area, and yield per land area in three time periods.***

average of 3.8% annually. That was a period when more growers began to overcome soil nutrient deficiencies with fertilizers. The gain in the 1970s over the 1960s averaged 1.9% per year. Since 1980, the average annual increases in yield have been a little less than 1.5%.

Even in developing countries where greatly increased productivity has already been achieved, recent gains are modest or negligible. Mexican farmers, for example, now obtain very good yields of wheat. During the latter half of the 1980s they reached a level of 4.2 metric tonnes per hectare. The growth curve has now flattened out, with no increase in average productivity thus far in the 1990s. Trade figures for 1995/96 show that Mexico (where Nobel Laureate Dr Norman Borlaug developed the 'miracle wheats' at CIMMYT) now imports wheat.

Global food security is clearly threatened by the likelihood of much slower gains in crop yields than experienced in recent decades, and limited land (Fig. 1.4) and water for expansion in farming. At least three strategies seem needed to assure food security in the next century:

1. A worldwide effort towards a more efficient food system through education, research, and adequate funding.
2. More effective programs, especially in the fast-growing developing countries, aimed at bringing birth rates down to the replacement level.
3. A more thorough realization, especially in the developed countries, that resources must be managed and used more efficiently.

Environmental Consequences of Increasing Demand for Plant Products

In 1992 the Union of Concerned Scientists issued a 'World Scientists Warning to Humanity' signed by some 1600 of the world's leading scientists, including 102 Nobel Prize winners. It notes that the continuation of destructive human activities 'may so alter the living world that it will be unable to sustain life in the manner that we know.' The scientists warned: 'A great change in our stewardship of the Earth and the life on it is required, if vast human misery is to be avoided and our global home on this planet is not to be irretrievably mutilated.' Environmental concerns raised included factors associated with population increases such as deforestation, topsoil erosion, fresh water availability and non-renewable resource depletion. Some major environmental issues associated with the increasing demand for plant products are introduced in the following section.

Loss of biodiversity

One concern over the loss of biodiversity arises from the extinction of plant species or genotypes. The practical concern is the loss of individual plant sources of genetic material to improve crop varieties and to provide new pharmaceuticals. Increased demands for land to grow crops will continually threaten natural ecosystems. Approximately one-half of the area of tropical forest lost each year is used as new area for crop production. The other half is used to replace worn-out, abandoned agricultural lands. Overall, the worldwide loss rate of tropical forests has been estimated to be about 1% per year for the period 1980–90.

A second important concern in the loss of biodiversity results from modern crop production systems of sole-cropping and monocropping. These practices mean that only genetically similar plants are grown together and that these plants are grown in the same fields year after year. (The historical developments of these practices in agriculture and their significance are discussed in detail in Chapters 2 and 3.) Therefore, large areas of crop plants, all potentially vulnerable to the same pests and diseases, are grown repetitively on much of the agricultural landscape. Chemical pesticides were introduced and successfully controlled many of these pests in the early years of high-yielding monocultures. Insects and fungi evolve, however, in these systems and the pests can eventually become resistant to specific pesticides.

Another problem of increasingly intensive agriculture is the increasing reliance on fewer varieties, that is, narrower genetic bases. (This issue is discussed in detail in Chapter 3.) This results because plant breeders continually search for wild plant types that would make crops more insect resistant, disease resistant, frost resistant, and with other tolerance traits. But there is a catch: to meet the growing demand for food for growing global populations, farmers have been quick to adopt more efficient, more uniform varieties. Farmers have also been clearing native habitats to create new farmlands on which to grow these high-yielding crop varieties. Unfortunately, this has commonly meant that old varieties and wild species of plants that may have important genetic traits for improving crop plants in the future, are disappearing.

A hope for counteracting the loss of genetic diversity in natural ecosystems is the generation of desired genetic traits using biotechnology. Unfortunately, biotechnology thus far has not produced any yield-raising technologies that will lead to substantial increases in maximum crop yields. (In Chapter 5 the limits to crop yields are discussed in detail.) Biotechnology, in the future, is expected to result in improved genetic resistance to pests, improved quality of product, and some improvements in postharvest qualities of our crops. Transgenic crops, as they are known, are a promising new tool for farmers to efficiently combat insects or weeds and to sustain yields while reducing the use of chemicals. (The issues associated with this technology are discussed in detail in Chapter 3.)

Land and soil degradation

Before the beginning of this century, almost all increases in agricultural production occurred as a result of increases in the area cultivated. By the end of the century, there will be no significant areas where agricultural production can be expanded simply by adding more land to production. To make matters worse, land is being lost to urbanization, industrial uses, and highways. In some heavily populated countries, including China and Japan, the amount of land available for food production is actually decreasing. For example, in the USA about 0.5 million hectares of productive soil are lost to development each year. Globally, the amount of crop land per person is falling at the rate of 1.5% per year (Fig. 1.5). Much of this lost agricultural land is now being compensated for by clearing forests, which in turn results in other environmental problems. On the southern edge of the Sahara, some 250,000 square miles of once-productive land have become desert over the past 50 years. As yields have failed to increase on existing crop lands, Africa's farmers have cut down forests or have moved on to even more fragile lands to survive. Africa loses roughly 5 million hectares of forest every year, primarily because of clearing for agriculture. Globally, factoring-in population growth and loss of arable land, the per capita arable land availability has declined to 0.24 ha per person, about half of what it was in the late 1950s.

With limited land area for crop production, crop yields per unit land area must be maximized. A problem with this approach, however, is the increased

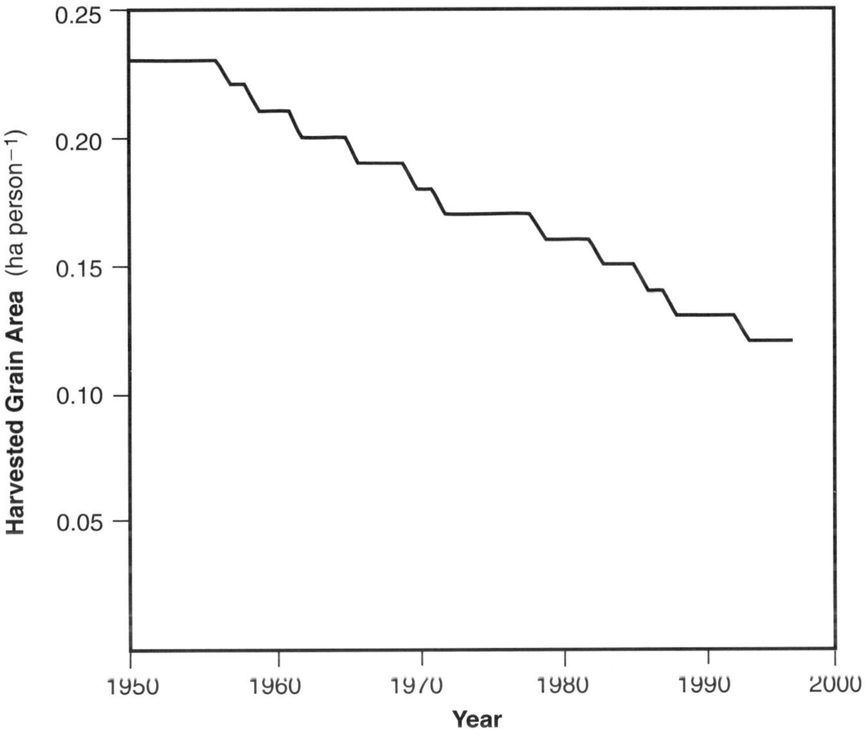

Figure 1.5 ***Crop land in production per person for the period since 1950.***

possibility of soil degradation. Soil erosion is a particular concern. For example, the average flow of the Mississippi River now dumps an average of 6.5 metric tonnes of eroded soil into the Gulf of Mexico *per second*.

The USA has lost one-third of its topsoil since Columbus set foot in the western hemisphere. Worldwide, approximately 10 million hectares of arable land are completely lost from crop production each year because of erosion. A particularly important feature of topsoil erosion is the loss of organic material that is in this layer of the soil. Declining soil organic matter influences several aspects of soils' productivity. These include reducing infiltration of rainfall, reduced water-holding capacity, decreased nutrient-holding capacity and decreased nutrient availability. All these problems result in decreased soil fertility. (These problems are discussed in detail in Chapter 6.)

Until recently, much of the loss of soil fertility could be compensated for by increases in the amounts of applied fertilizer. For example, one of the soil nutrient constraints to yield is phosphorus, especially in the impoverished soils of the tropics. The use of phosphate as fertilizer had been rising faster than the population growth rate. The US Bureau of Mines believes there are about 34 billion metric tonnes of phosphate rock left to be quarried. At the present rate of consumption, this will run out by about 2050.

Fertilizer use in technologically advanced regions such as Europe, the USA and Japan has already stopped increasing and is actually declining in some areas. Currently, developing countries are applying fertilizer at about 90% of the amount per unit area of high income countries. This is a substantial change since the early days of the Green Revolution, when fertilizer use in the developing countries was less than 20% of that of the high income countries. The high levels of fertilizer use worldwide result in another set of environmental problems associated with nutrient leaching and runoff from plant production systems.

Water

Despite 71% of the Earth's surface being water, only 3% is fresh water and half of that is in polar ice caps. Yet, humanity's use of fresh water has quadrupled since World War II. There simply is not enough fresh water to quadruple use again.

Water shortages and conflicts over water threaten to become more serious in the future. For example, water shortages in much of northern and western China will be a factor in China's agricultural and economic growth. In 1992, the Yellow River was dry at its mouth for more than 3 months due to a combination of use and weather. In 1995, 622 km of the Yellow River again went dry and in 1996, sections of the great river went dry. The aquifers that supply 50% of Beijing's water are falling at the rate of a meter a year (40 m since 1950). This is causing the ground level to sink as groundwater is extracted.

In many regions, plant production systems are viewed as the prime sources of the fresh water for urban and industrial purposes. This is particularly true where there are large areas of forests or grasslands, although watersheds with large areas of intensive crop production are also very important in many regions because of increasing demands for clean water. Therefore, there is considerable concern about the use of pesticides and mineral nutrients in plant production systems. These chemicals may eventually appear in the urban water sources derived from the runoff and deep percolation of plant production systems. These concerns relate directly to the intensity and type of management imposed in plant production.

In addition, plant production systems themselves need huge quantities of water. (The details of water use in plant production are discussed in Chapter 7.) In drier regions the need for water is supplied, to varying degrees, by irrigation. In most areas where irrigation needs are supplied by groundwater, e.g. western Great Plains in the USA, water tables are being lowered because withdrawal rates exceed recharge rates. Only about 75% of the water removed is being recharged, so that there is a net loss to the groundwater of 25% of what is being used. Clearly, this continual decrease in stored water cannot persist as the cost for obtaining the limited amount of remaining water precludes plant production.

When industry or urban areas use water, most of the water is returned to the environment as a liquid, although it may not be as clean as it was originally. When water is used in plant production systems much of the water is evaporated to the atmosphere and it is transported away from the water-deficit location. Consequently, plant production systems commonly account for the great bulk of net losses of water from localized regions.

Waterlogging and salting are also threats to productivity of irrigated land. When river water is diverted on to the land for irrigation, part of the water percolates downwards, sometimes raising water tables. If the water table rises to within a meter of the surface, deep-rooted crops suffer. When the water table gets within a few centimeters of the surface, water rapidly evaporates from the soil, leaving an increased concentration of salt at the soil surface. Unless an underground drainage system is installed to lower the water table, the accumulating salt eventually affects the fertility of the land, as happened with early Middle East civilizations. Excessive salinity is estimated to have affected the productivity of some 10–12% of the world's irrigated lands.

Weather and climate

Plants are extremely sensitive to weather and climate. The prevailing weather patterns and summary of these patterns into discernible climatic zones are important features in assessing the potential success of plant productions systems. Factors such as latitude, geography, and topography all play important roles in determining weather patterns and climate. Such background considerations are important in examining the environmental issues associated with plant production systems. (Chapter 10 discusses weather and climate in detail.)

In addition to rainfall, two very important aspects of the weather and climate concerning plant production systems are light and temperature. Environmental variations associated with changes in either of these variables can greatly alter the productivity of plant systems.

Light, technically called solar radiation, is the basic energy source for plant growth. The amount of plant growth is closely related to the total amount of solar radiation received by the plants. Therefore, changing atmospheric conditions such as cloudiness can influence plant production. In addition, specific wavelengths of solar radiation might have large influences on plants. For example, there is concern that decreases in atmospheric ozone allowing greater penetration of ultraviolet light to the Earth's surface might greatly damage plants. (These concerns are discussed in detail in Chapter 8.)

Temperature is one of the most important environmental factors that influence plant production systems. Certainly, extreme temperatures in an area set hard limits to the types of plants that can be grown and to the time of year when they may be grown. Cool and freezing temperatures are particularly important in restricting plant growth. In addition, the pace of plant development is commonly closely associated with temperature. Warmer temperatures, up to an optimum, result in more rapid plant development. There can be both positive and negative aspects of changes in the rate of plant development. These responses of plants to temperature are important in evaluating the effect of weather and climate on plant production systems. (The influence of temperature on plants is discussed in detail in Chapter 9.)

Until recently, it was assumed that the climate and weather for specific regions were fairly stable over the long term. As the old saying goes, 'Everybody talks about the weather, but nobody does anything about it.' This assumption is now realized to be not true and humans collectively are having an influence on the weather. A by-product of the energy-consuming lifestyle especially prevalent in the more affluent countries, is the burning of fossil fuels that releases carbon dioxide and other gases into the atmosphere. It is now known that the increasing emissions of these gases are having an influence on the global environment.

The steady increases in carbon dioxide have a direct influence on the rate of photosynthesis and growth of many plant species. Therefore, some species are expected to have substantially increased growth rate while others will have little change. These differences in response to carbon dioxide will influence the relative advantage among plant species in both natural and plant production ecosystems.

Besides the direct effect of carbon dioxide, carbon dioxide and other gases can act as 'greenhouse' gases in the atmosphere. These gases absorb the thermal-infrared radiation radiated from the Earth's surface, and this absorbed energy results in a heating of the atmosphere. Increasing the concentration of these gases in the atmosphere is predicted to increase the temperature of both the atmosphere and the global climate. In late 1995, the International Panel on Climate Change concluded that 'the balance of evidence ... suggests a discernible human influence on global climate.' There remains considerable uncertainty about how much and how fast climate will change. Even in the most conservative case the average rate of warming is expected to be 'greater than any seen in the last 10,000 years,' a major change in climatological terms. (This problem in terms of plant production systems is discussed in detail in Chapter 11.)

Summary

The human population has increased at an unprecedented rate with the current population at 5.9 billion people. Increasing numbers of people have resulted in tremendous demands on plant production systems for increasing output of food, fiber, and fuel. Early in the need to meet these demands, there were substantial infringements on natural ecosystems to convert them to plant production systems. Land was cleared for food crops and cotton, and natural forests were replanted with managed regrowth tree plantations. In recent times, when the population increase has itself greatly increased, most of these demands have been met mainly by increasing the productivity per unit land area.

The greatly increased demands for food, fiber, and fuel have many important general consequences. World trade markets can be readily destabilized with production shortages in specific areas of the world. This economic sensitivity can be especially significant in developing regions of the world where food production systems may be poorly developed. The economic or political disparity in food supplies in particular, can greatly influence the health and well-being of many people in developing countries. Children especially are vulnerable to a lack of food supplies. Finally, there is the important question of whether it is possible to sustain recent increases in plant productivity to meet the demands of increasing human population.

The increased demands for output from plant production systems are associated with many critical environmental issues. The environmental factors involved in these issues include biodiversity in the management and genetic selection of plants, soil fertility and conservation, water use and irrigation, and weather and climatic variables. Background information to evaluate these various environmental issues associated with plant production is presented in the subsequent chapters of this text.

Further Reading

Brown, L.R. and Kane, H (1994) *Full House: Reassessing the Earth's Population Carrying Capacity*. W.W. Norton & Company, New York.

Evans, L.T. (1993) *Crop Evolution, Adaptation and Yield*. Cambridge University Press, Cambridge, UK.

Gardner, G.T. (1996) *Shrinking Fields: Cropland Loss in a World of Eight Billion*. Worldwatch Institute, Washington, DC.

Johnson, H.J. (1996) Food Security in Africa. *Testimony before the Subcommittee on African Affairs, Committee on Foreign Relation*, US Senate. 23 July 1996. US Government Accounting Office, Document #GAO/T-NSIAD-96-217.

Kegley, C.W., Jr and Wittkopf, E.R. (1995) *World Politics, Trend and Transformation*. St Martin's Press, New York.

Mazur, L.A. (ed.) (1994) *Beyond the Numbers*. Island Press, Washington, DC.

Moffett, G.D. (1994) *Critical Masses: the Global Population Challenge*. Viking Press, New York.

Ecological Perspective in Plant Production

2

F.P. GARDNER AND T.R. SINCLAIR

Ecology is the study of the physical and biological interrelationships of communities of organisms. Each plant and animal in a community may influence the life patterns of all other members of the community, consequently, a very complex and dynamic situation exists. Even in the simplest community, such as those generally associated with plant production systems, the number and diversity of organisms is large. The myriads of interactions among these organisms result in a dynamic and changing community structure. The great diversity in natural biological communities allows organisms to exploit temporally and spatially every conceivable niche of the environment. Study of these communities embraces a multitude of biological and physical scientific disciplines. Integration of the principles from each of these disciplines leads to another higher level discipline, called ecology.

Ecosystems

A community of organisms with a fairly uniform composition that share a similar physical environment, conceptually defines an ecosystem. Clearly, the definition of an ecosystem is arbitrary and the dynamic status of ecosystems challenges any attempts to characterize the community of organisms. Nevertheless, it is conceptually useful to identify traits associated with specific ecosystems such as rainforests, tropical rangelands, or croplands.

Natural ecosystems are generally highly complex, involving perhaps hundreds of species. Many species in the community are primary producers and fix carbon dioxide into organic compounds via photosynthesis, while others are consumers of organic materials in various food chains and webs. As illustrated in Fig. 2.1, biotic components of the community include producers (plants), consumers (herbivores, carnivores, omnivores, predators), and scavengers and decomposers (detrivores and saprovores). The biotic components interact with the physical environment so that the flows of energy and mass are altered.

CAB INTERNATIONAL 1998. *Principles of Ecology in Plant Production*
(eds T.R. Sinclair and F.P. Gardner)

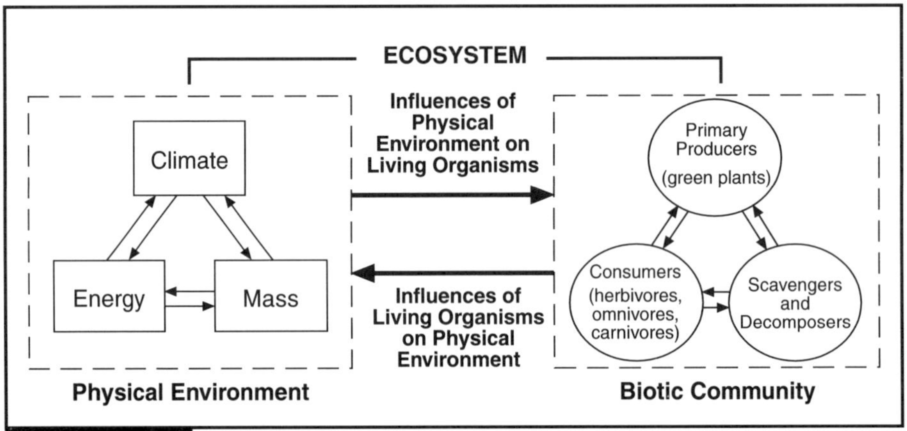

Figure 2.1 ***Schematic of the physical environment and biotic community that describe an ecosystem*** *(redrawn from Cox and Atkins, 1979).*

The study of the flows of energy and, particularly, of mass within an ecosystem is an important approach to characterizing an ecosystem. Within an ecosystem, important processes include primary production, consumption, and decomposition interacting with the abiotic factors, resulting in energy flow and nutrient cycling. Since ecosystems generally are space limited, exports of mass and energy are also important when consumers move to other environments. One of the important features of plant production systems is that exports are often massive as the harvested plant or animal product is removed from the system. In crop production, for example, maize grain produced in the ecosystem of the US Cornbelt is consumed in western feedlots, southeastern dairy herds, or in beverages everywhere. In addition, roughly a fourth of the crop is exported to other countries.

The ecosystem concept is best understood perhaps by visualizing the largest and definable ecosystem known as a biome. A classic example is the tropical rainforests, which thrive on the 2000 km belt of land on each side of the Equator. The tropical climate that favors development of a rainforest has a mean temperature of 26°C (79°F), a nearly constant daylength of 12 h, and a rainy season of 10–12 months. Survival of this natural terrestrial ecosystem is threatened due to rapid commercial and agricultural development. That is, the export of materials from this ecosystem far outstrips its ability to regenerate the various components of the nutrient and mass cycles.

The exploitation of the tropical rainforests as a raw plant production ecosystem has dire consequences because of the rainforest's biological diversity and enormous size. Tropical rainforests consist of tens of thousands of plant and animal species, reportedly more than 50% of the plant and animal species on Earth. Countless organisms have a niche at various levels in the forest canopies and rhizospheres (root zone). A uniform leaf canopy, as seen from above, disguises the large number of species that live in the rainforest. There are many layers within the canopy that offer a niche for a wide diversity of plants. Aerial plants (epiphytes) grow from tree branches and trunks. Very large to small lianas or vines climb and twist around tree trunks and branches. The trees and lianas filter out nearly all solar radiation so the forest floor is low in light and only extremely shade-tolerant species survive on the floor. The effect is the appearance of tunnels in the dark-forest canopy. Animal life in this unique habitat is equally varied and complex, ranging from primates to microscopic animals. The rainforests' consumers often include humans, for example, the Pygmies of the tropical forests of Zaire and the Indian aborigines of the Amazon.

Structure of Ecosystems

The structures of natural ecosystems are taxonomically and physically diverse, especially as they near maturity and a climax state of development. The climax state is conceptually visualized as a steady state condition. Theoretically, the number and kinds of species and production reach a near equilibrium under the prevailing dominant climate and edaphic factors. Of course, the dynamic nature of the elements in the ecosystem community prevents a true steady state from developing as the various competing organisms protect and expand their niches. Population densities vary with weather and predation. Evolutionary changes occur as organisms adapt to changing weather and environment.

Plant production ecosystems are plant and animal communities designed and controlled to meet human objectives. To facilitate the production of the plant components desired by humans, an attempt is made to impose constraints on unwanted organisms that may flourish in a plant production ecosystem. Consequently, plant production ecosystems are relatively simple, when contrasted with natural ecosystems. Modern plant production ecosystems in agriculture and forestry are often based on a single genetically uniform species grown at high population densities. These ecosystems are commonly established with the goal to maximize the yield of the harvested component.

In practice, the desire of humans to maintain a simple ecosystem almost invariably conflicts with the exploitation of various open niches by undesirable organisms. This exploitation of niches by differing organisms, and the resultant pressure towards diversity of organisms in plant production ecosystems, results in an intrusion by unwanted organisms (weeds, insects, and microorganisms). The unwanted organisms compete with the desired organism for available resources or they may seek to consume the crop species. The desire of humans to maintain simple ecosystems against the pressures of opportunistic organisms causes humans to use various methods, including chemical and biological intervention, to arrest the effects of these unwanted invaders.

Agricultural ecosystems, including crops and grasslands, occupy about 1.4 billion hectares (Table 2.1), or slightly over 10% of the ice-free land. Contrasted with natural ecosystems, they are populated with relatively few species, often a single species, if pests and soil biodata are excluded. The number of species usually varies inversely with management intensity, except in urban areas.

Table 2.1. Estimated land area of the major natural ecosystems and crop lands on earth (adapted from Cox and Atkins, 1979).

Ecosystem	Area (10^9 ha)	t ha^{-1} $year^{-1}$
Tropical forest	2.0	25
Temperate forest	1.8	12
Desert scrub	1.8	0.8
Savanna	1.8	10
Crop agriculture	1.4	10
Boreal forest	1.2	3
Temperate grassland	0.9	5.5
Arctic and alpine tundra	0.8	1.5
Woodland and scrub	0.7	6.5

Commercial agriculture in developed nations relies heavily on monocultures of 'tailored' plant varieties that leave many ecological niches for invaders (weeds, insects, and diseases). Chemical, biological, and/or management controls are invariably necessary to sustain yield in monocultures.

Grasslands are a departure from the monoculture model in that two or more plant species are planned and the species are usually perennial. Grassland food chains consist of forage species, herbivorous animals (including livestock, rodents, and insects), and carnivores, including birds. Biological diversity is increased compared with a monoculture and the potential for pest epidemics is decreased.

Plant production ecosystems of various structures also merge with natural ecosystems that can result in the formation of patterns of biological diversity. Forests of differing ages and species composition introduce diversity into the ecosystems. Fields of cultivated crops, meadows, or pastures interspersed with woodlots, a common occurrence in Europe and northeastern and southeastern USA, is an example of such patterns. This diversity has the advantage of less pest susceptibility as compared with continuous monoculture. However, diversity is usually sacrificed as cultivated crops replace natural ecosystems, farm size and field sizes increase to accommodate larger machinery, and crop rotations are eliminated. The wheat rust (*Puccinia graminis*) of the Great Plains, USA in the 1930s and the Irish potato blight (*Phytophthora infestans*) in the 1840s are grim testimonies to the consequences of large land areas devoted to monocultures of genetically uniform crops.

Village agriculture, common to Africa and other developing countries, uses small areas of multiple cropping (10 or more species), and is more akin to natural ecosystems than modern farming. Like natural ecosystems, these agricultural ecosystems also rely largely on recycling plant nutrients and on natural pest control by species barriers and immunities.

Perturbation

Perturbation (destabilization or disturbance) of ecosystems, at least from natural causes, is a process in ecosystem development. Fire greatly influences the development of natural ecosystems and is an essential feature of some. For example, California chaparral requires fire for propagation by breaking seed dormancy. Fire also controls dicot weedy trees in coniferous forest, which may in time dominate coniferous forests.

In 1988, fires burned thousands of hectares in Yellowstone National Park, USA. After considerable debate, the fires were deemed to be natural phenomena and, therefore, allowed to burn unimpeded. This decision was controversial even among scientists. Some considered that the result was wanton destruction of forests, wildlife, and an aesthetic and recreational area. Also, the fire resulted from human activity rather than from a natural cause. All agreed that fire is one of the natural forces in the development of this ecosystem, but this fire was an extreme case. To what extent can the fire be viewed as a natural phenomenon in contrast to a destructive human act, and what will be the long-range consequence?

Perturbations also occur in plant production ecosystems by such practices as animal grazing and soil tillage. The perturbation of the tall- and short grass prairies followed the invention of the steel moldboard plow. A very different vegetation on the resultant croplands replaced the native species. The tall-grass prairie in the whole of north central USA essentially disappeared in about 50 years. The prairie's legacy is the vast fertile area left by the tall grasses, now referred to as the Corn and Soybean Belt. After 1900, the short-grass prairie ecosystem, currently the Wheat Belt of the USA, was similarly converted to a plant production ecosystem.

Ecological Succession and Ecotypes

If an ecosystem is altered by perturbations, its stable component species or ecotypes may be supplanted by new, pioneer vegetation. The vegetation found on strip-mining spoils, cutover forest, and abandoned cropland differs from the natural ecosystem in species number and type. Plant diversity and value as a habitat for a diversity of animal species is very small in pioneer vegetation. Following the pioneer stage, the increase in species number is essentially linear over time until near equilibrium conditions again develop. At this point, species ingress equals egress in response to the changing environment.

Observations of plant succession on strip-mine spoils in eastern Ohio, USA showed that the legumes white clover and black medic were among the earliest pioneers to colonize the limestone or shale spoils. These spoils consisted of fine decomposing rock particles that were high in cations but extremely low in organic matter and nitrogen. Evidently, these legumes emerged from hard seed transported to the area by birds. Weedy species with well-adapted mechanisms for wind distribution, such as thistles and dandelion, invaded next. Brush followed weeds, and in a short time, without pressure from livestock grazing, a broadleaf temperate-forest ecosystem emerged. At maturity (climax), the dominant species in the ecosystem on well-drained sites of this area was oak–hickory forest. The climax ecosystem in areas where grazing was imposed consisted of temperate grasses (Kentucky bluegrass was usually the dominant species), white clover, weeds, and a population of bushes and trees.

Natural succession on lands abandoned for cropping in central Oklahoma and southeast Kansas, USA consisted of four stages of invasion: (i) pioneer weeds (*Ambrosia psilostachya*, *Helianthus annuus*, *Chenopodium album*, and *Sorghum halepense*); (ii) an annual grass (triple awngrass); (iii) a perennial bunchgrass (little bluestem); and finally (iv) the true prairie (little bluestem, big bluestem, switchgrass, and indiangrass). The succession pattern was altered by the facts that triple awngrass is highly tolerant to low fertility and that some weeds produce chemical toxins that inhibited growth of other plants (allelopathy). Low-fertility tolerance and allelopathy probably play important roles in many plant successions and the development of natural ecosystems. It takes about 40 years to complete the succession from abandoned cropland back to true, natural prairie.

As ecosystems mature, plant species that have superior adaptation to the ecosystem become dominant. These adapted plants, which have evolved to fill these niches, are called ecotypes. Ecotypes from a grazed sub-climax ecosystem have found important places in agriculture. Examples of ecotypes of the common alfalfas are Kansas common and California common alfalfa, which have taken the name of their states of origin. Similarly, Empire is an ecotype of birdsfoot trefoil (a permanent pasture legume) selected from an overgrazed pasture in New York State. This ecotype now has substantial significance in agriculture. Lincoln, Achenbach, and Southland bromegrass are ecotypes selected from closely grazed pastures in the Great Plains states.

Food Chains and Webs

A food chain is a trophic (nutritional) structure (producers–consumers–decomposers) used to describe the flow of mass and energy in the ecosystem (Fig. 2.1). The original source of nearly all energy of an ecosystem is the solar energy absorbed by green plants, which are identified as the primary producers. The next trophic level is formed by the consumers of the primary producers. Living plant materials are consumed directly by various organisms, and decomposers consume senesced plant components. The herbivores and omnivores may be consumed directly by carnivores, or with the death of the animals they become the feedstock for various scavengers and decomposers. As energy passes from one trophic level to another in the food chain, up to 90% of the energy is dissipated.

The trophic pathway of energy in a simple plant production ecosystem is illustrated in the production of meat for human consumption (Fig. 2.2). Initially, only a small percentage of solar energy (1–3%) is absorbed by the forage crop.

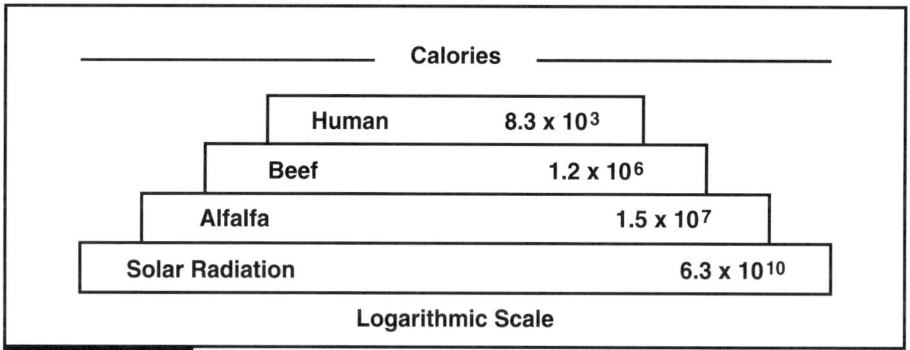

Figure 2.2 ***Pyramid of energy amounts that flow from the solar energy received at the earth's surface to that in a human body. Note that the scale is logarithmic to allow a representation of the orders of magnitude change in energy.***

In the next step, the beef animals convert only 5% or less of the forage plant to meat energy. Finally, about 5% of the meat energy when consumed is retained in the human body. This example illustrates that removal of mass from the ecosystem in the form of animal products can only be sustained with large energy inputs from the sun.

Materials that are not metabolized and dispersed in the ecosystem at one trophic level are passed to the next trophic level, and consequently are concentrated. Concentrations of some materials can reach toxic levels in foods consumed by carnivores and omnivores, including humans due to their position in the food chain. These materials include chlorinated cyclic hydrocarbons, such as dichlorodiphenyltrichloroethane (DDT), which are deposited in animal fat, and certain heavy metals, such as mercury and lead, which accumulate in bone marrow. The concentrations of these materials at the next trophic level produce the effect termed 'biological magnification'.

Primary Production and Consumption

The use of solar energy to accumulate chemical energy as biomass by green plants is important in all ecosystems. Energy flow in an ecosystem is dependent on the ability of these primary producers to absorb solar energy and to chemically trap the energy into plant constituents. However, the actual amount of primary production by ecosystems is dependent on the climate. The most productive natural ecosystem is the tropical forest, which is humid and constantly warm (12-month growing season), and the least productive natural ecosystems are the desert and tundra (Table 2.1). The most productive ecosystems managed by humans for plant production are typically agricultural crops to which humans have added large amounts of outside resources of nutrients and water. Examples of such ecosystems are annual grain crops, such as maize and rice.

Consumption of primary production by heterotrophs is easily observed in natural and grazing ecosystems. However, in commercial farming the products of primary production often are not consumed *on the farm*. In the US, maize may be consumed in large western feedlots or as sugar in soft drinks. Further, the product may not be ingested by any other organism as happens in the production of fiber and wood.

Nutrient Cycling

In contrast to energy, which generally flows in only one direction between trophic levels, movement of minerals in food chains is bidirectional or cyclic. In

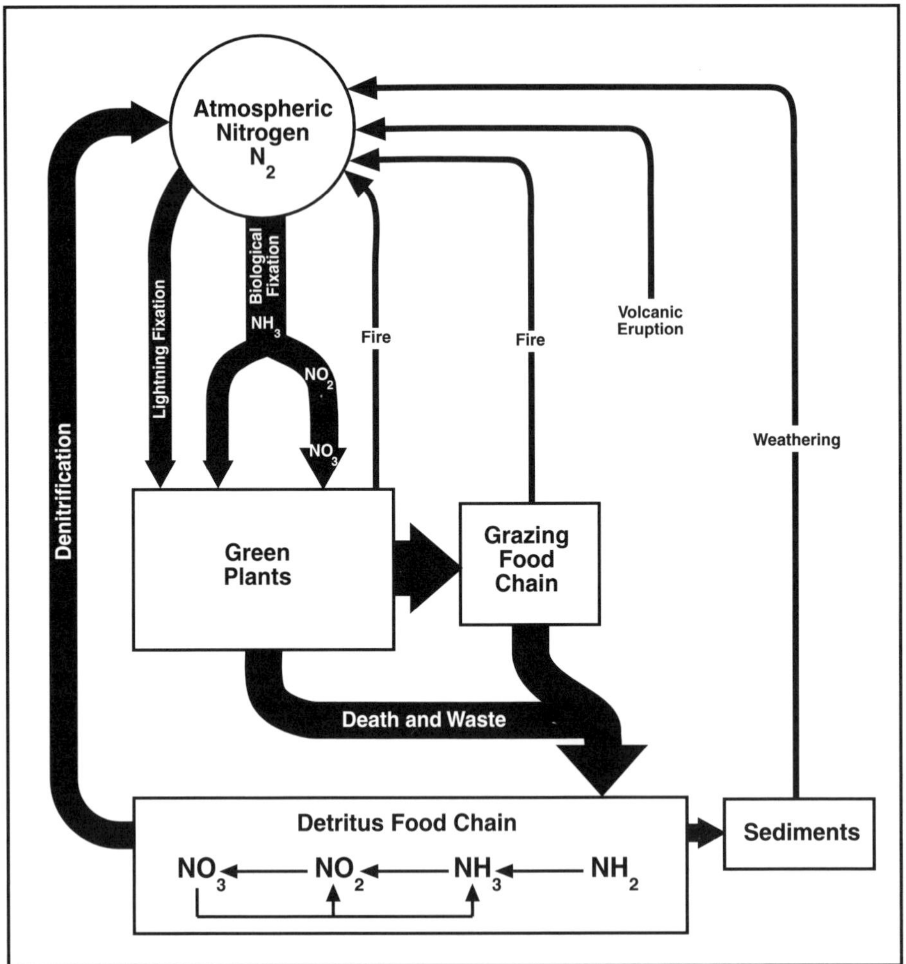

Figure 2.3 ***Schematic of nitrogen cycle in a natural ecosystem. The size of the arrows reflect the size of the fluxes in each process*** *(Cox and Atkins, 1979).*

time, minerals or nutrients cycle back to their source pools so they can be reused. As an example, the nitrogen cycle is illustrated in Fig. 2.3. In natural ecosystems, nitrogen flux through the various compartments is at equilibrium. In plant production ecosystems, in contrast, nutrient levels may decline due to bio-oxidation of organic matter, erosion, leaching, and nutrient removal in harvested products. In both ecosystems, organic waste (plant, animal, bacteria) is mineralized in the detritus food chain by heterotrophs. Nitrogen fixed by lightning (10–15 kg ha^{-1} $year^{-1}$) can accumulate in biomass. Specific microbes are critical in biological nitrogen fixation by symbiosis or association with plants and by free-living microorganisms. Biological nitrogen fixation is vitally important in natural ecosystems and plays an important role in many plant production ecosystems. Microbes are also essential in the transformation of the reduced form of N (ammoniacal) to the oxidized form (nitrate), generally the most easily available form for most terrestrial plants.

Comparison of Natural and Plant Production Ecosystems

The above discussions have alluded to important differences between natural ecosystems and those managed by humans for plant production. It is important to recognize explicitly some important features that differentiate a plant production ecosystem from a natural ecosystem.

1. *Controls*. Natural ecosystems are controlled internally by feedback mechanisms; plant production ecosystems are controlled externally, and are human directed by economic and social factors.
2. *Energy flow*. Both systems depend on solar energy to form chemical energy as a result of photosynthesis, and to increase plant mass. In addition, most plant production ecosystems require external energy subsidies from fossil fuels in the use of machinery, fertilizers, and pesticides to increase photosynthetic productivity. Discrete food chains are identifiable in grazing systems but may not exist in other plant production systems. Consumers of plant products are generally far beyond its boundaries, e.g. lumber exported to Japan or soybean meal shipped to Europe.
3. *Diversity*. Natural ecosystems are biologically diverse, often containing 300 or more species; plant production ecosystems are commonly monocultures that attempt to exclude most competing organisms.
4. *Litter (detritus)*. Natural ecosystems, depending on species, climate, and growth rate, deposit large amounts of litter on the soil. Litter feeds the detritus food chain, releasing plant nutrients for recycling and reuse by the autotrophs. Plant production ecosystems in many cases leave less litter on the soil, and sometimes, plant residues are completely removed or burned. If plant residues are left on the land, they can be rapidly bio-oxidized because of the management required in plant production.
5. *Nutrient cycling*. Natural ecosystems recycle nutrients efficiently via the detritus food chain. Nutrients stored in the biomass are also recycled within the

plant. Nutrients move from old to new plant parts once the initial plant structure has been established, resulting in an effective and efficient nutrient trap. On the other hand, plant production ecosystems commonly require large inputs of nutrients because considerable amounts of nutrients are removed in the harvested plant components. Some cycling of nutrients occurs with crop plants, but this process is not relied upon to a large extent in modern systems.

6. *Soil organic matter and tilth.* Natural ecosystems tend to accumulate organic matter and improve soil tilth, while increased soil aeration as result of tillage in plant production ecosystems can result in bio-oxidation of organic matter.

Sustainability of ecosystems

Ecosystem sustainability has become an important issue underlying many economic and political decisions regarding plant production and the environment. By definition, natural ecosystems are inherently sustainable as they persist in a dynamic equilibrium. Even with perturbation, an ecosystem, given sufficient time, commonly recovers to near its original state. Although, cataclysmic perturbations such as volcanoes or floods might sufficiently alter the local environment that a new ecosystem will develop.

Plant production ecosystems are not sustainable without human intervention. Invariably at some point in the life cycle of the plant production ecosystem outside energy must be expended in maintaining the ecosystem. Even in regrowth forestry where human intervention is small, forests must be harvested and techniques to insure regrowth are required. In agricultural and horticultural systems the level of human intervention is high. The production ecosystem is sustainable only as long as the intervention continues.

One of the first issues in resolving the sustainability of a plant production ecosystem is whether the reward to humans is sufficient to continue the effort in maintaining the ecosystem. If agriculture is at the subsistence level, humans have little option but to expend full effort to sustain the ecosystem. Although one important approach, when the population density has been low, was to manage a compromise between a natural and plant production ecosystem. A small section of land would be cleared in the forest to support crop production. With time, the nutrients would be ‘mined’ from the soil, and weed seeds and crop pathogens would invade and accumulate in the clearing. As a result, the clearing would eventually be abandoned and the natural succession process would be allowed to proceed toward the original natural ecosystem. A new section of the forest would be cleared and used for cropping until it too was eventually allowed to return to the natural state. In cases where there has been human habitation for centuries, it is likely that the same land cycles between short periods in cropping and long periods in natural state.

In addition to the issue of human reward in judging the sustainability of a plant production ecosystem, the system can only be sustained if energy and other resources are readily available. Energy is required to till the soil and to protect the crop from invading weeds, insects, and pathogens. After the inevitable depletion of resources in plant production ecosystems – mineral nutrient losses from the soil are among the most important – these resources must be replenished to maintain plant productivity. For example, mineral resources must be available and the capability must exist to apply them back to the soil. In many subsistence farming areas a major problem is the lack of infrastructure and economic framework for supporting high levels of crop fertilization to sustain high crop yields.

Finally, sustainability of the plant production ecosystem requires that the environment is not degraded to the extent that cropping becomes permanently unproductive. Erosion and degradation of the soil have occurred in many plant production ecosystems. This degradation can readily result in a decrease in yields to levels where the production (or 'rewards') is too low to allow the ecosystem to be sustainable. The accumulation of salts on irrigated lands is another example of degradation that results in very low crop yields and the plant production ecosystem is no longer sustainable. Maintaining favorable environments for a sustainable plant production ecosystem is a great challenge in providing food and fiber for the world's population.

Summary

Ecology is the study of organisms interacting with their biological and physical environments. The ecosystem concept of mass and energy flow in the biological framework of food chains/webs (primary producers, consumers, and decomposers), is a basic principle in the study of ecology. Natural ecosystems are controlled by internal feedback mechanisms, including nutrient cycling and biological diversity that buffer the system. Plant production ecosystems are generally monocultures controlled by external forces that normally require large external inputs, usually derived from fossil fuels to protect them against pests and nutrient deficiency. Natural ecosystems experience perturbations from fire and human activities. Plant production ecosystems can be virtually destroyed by pests that can easily invade niches in the genetically uniform structure of the agricultural species. The sustainability of the plant production ecosystem depends on maintaining the environment so that the rewards harvested from the ecosystem remain at a high level.

Further Reading

Altieri, M.A. (1994) *Biodiversity and Pest Management in Agroecosystems*. Food Products Press, New York.

Carroll, C.R., Vandermeer, J.H. and Rosset, P. (1990) *Agroecology*. McGraw-Hill, New York.

Coleman, D.C., Cole, C.V. and Elliott, E.F. (1984) Decomposition, organic matter turnover, and nutrient dynamics in agroecosystems. In: Lowrance, R., Stinner, B.R.

and House, G.J. (eds) *Agricultural Ecosystems*. John Wiley and Sons, New York, pp. 84–104.

Cox, G.W. and Atkins, M.D. (1979) *Agricultural Ecology, an Analysis of World Food Production Systems*. W.H. Freeman and Co., San Francisco.

Haas, H.J., Evans, C.E. and Miles, E.F. (1957) *Nitrogen and Carbon Changes in Great Plains Soils as Influenced by Cropping and Soil Treatments*. Technical Bulletin No. 1164, USDA, Washington, DC.

Jacobs, M. (1988) *The Tropical Rain Forest*. Springer Verlag, New York.

Loomis, R.S. and Connor, D.J. (1992) *Crop Ecology: Productivity and Management in Agricultural Systems*. Cambridge University Press, Cambridge.

Lowrance, R., Stinner, B.R. and House, G.J. (eds) (1984) *Agricultural Ecosystems*. John Wiley and Sons, New York.

Rice, E.L. (1974) *Alleopathy*. Academic Press, New York.

Tomanck, G.W., Albertson, F.W. and Riegel, A. (1955) Natural revegetation on a field abandoned for thirty-three years in central Kansas. *Ecology* 36, 407–412.

Vasey, D.E. (1992) *An Ecological History of Agriculture*. Iowa State University Press, Ames, Iowa.

Diversity and Genetics in the Development of Crop Species

3

D.A. KNAUFT AND F.P. GARDNER

Origins of Agriculture

About 10,000–12,000 years ago an important transition took place in the way people lived. Before this time, food was obtained by hunting wild animals and gathering food from plants in natural communities. In the transition to agricultural ecosystems, plants and animals were genetically adapted by human selection, that is, domesticated. Selected plants and animals were allowed to increase in numbers while the numbers of unwanted organisms were decreased. Special techniques of sowing, hoeing, irrigating, and fertilizing were developed to provide not only greater quantities of food, but a more stable supply.

It is not known when this process of domestication took place. Evidence cannot be traced with certainty since the process was evolutionary and humans moved with their domesticated species to new settlements (shifting cultivation). Fortunately, earlier peoples were untidy and many discarded items have survived to this day to help in understanding the origins of agriculture. These remnants include carbonized pieces of cereal grains and tools made of stone, such as scythes for harvest of grain and mortars and pestles for grinding of food. Ancient origins of other seed, including squashes and beans, are present, as well as the discarded remains of pottery.

Archaeologists have collected and tentatively dated these pieces of evidence. They have learned that early agriculture began independently in several places around the world. It is possible people were cultivating barley nearly 18,000 years ago in southern Egypt. Several archaeological sites in Mesopotamia over 10,000 years old have produced evidence of tools used in agricultural activities, and carbonized remains of wheat and barley. In southeast Asia, roots and vegetables were probably domesticated either simultaneously, or even earlier than plant domestication in the Fertile Crescent of Mesopotamia. Later, it appears that maize, beans, gourds, and other plants were cultivated in Central America. Still later, other plants, such as potato, cassava, and sweet potato were domesticated in South America.

 Principles of Ecology in Plant Production (eds T.R. Sinclair and F.P. Gardner)

The impetus for people to begin the domestication of plant species remains a puzzle. It seems unlikely that, given equal choices, people would choose the more rigorous and monotonous life of tilling the soil and tending crops, as compared with hunting and gathering. Nor is it likely that agriculture developed when people faced a scarcity of food. The time required for the evolution of agriculture would have been too long to provide people with adequate sustenance in a time of crisis. It is more likely that agriculture may have developed in situations where hunting and gathering was especially successful. Success in hunting and gathering would have allowed the human population in a settlement to increase. With increasing population pressures, there may have been a need to supplement and stabilize the food supply with harvests from cultivated plants. The plants selected for cultivation to augment the food supply, would be likely to have come from the seeds gathered from the most desired individual plants in the natural ecosystem.

The seed selected for the cultivated crop would have been the first step in plant domestication and agriculture. Gradually, the techniques needed for plant domestication could have developed. As populations increased further and settlements became larger, there is likely to have been increased pressures for greater food supplies within a short distance of settlements. Over time, the need for food for an increasing population could have resulted in increasing demand for cropping selected plant species in close proximity to settlements. The gradual shift over time to a greater reliance on cultivated plants could have set the framework for crop domestication.

Spread of Crop Species

From the various areas where farming originated, the practice appears to have spread throughout much of the world. In Europe, agriculture spread after the last Ice Age when people familiar with farming moved into the area from Turkey, Greece and Mesopotamia. River valleys contained deep, rich soil laid down during the previous Ice Age. Humans occupied these valleys, bringing along with them their farming techniques or developing them as they settled. The people pushed up the Danube and settled Hungary, Austria, and Germany. Other groups came through Spain to France and Britain. Many crops were domesticated by 5000 BCE (before the Christian era) including several types of wheat, beans, barley, and flax.

Agriculture also spread south from the Fertile Crescent region. Here a different type of agriculture had to develop, since rainfall was more scarce and less predictable than in Europe. Empires such as the one established in Babylon relied on irrigation systems for adequate food production. Farming in the Nile Delta region of Egypt may have moved west along the northern coast of Africa by 4000 BCE. It is not known why agriculture did not readily spread south into Africa until more recent times. The great barrier of the Sahara Desert and the plentiful supplies of food in the humid tropics may have discouraged the movement of agriculture into central Africa.

Farming moved east from the Fertile Crescent into northwestern India by 3500 BCE. There is also evidence that plants domesticated in Africa were brought to the region. Other agricultural activities in India appear to trace their origins to regions of southeast Asia. Agriculture in Southeast Asia spread throughout many parts of the world, including the islands of the Pacific Ocean. Although agriculture in Asia was probably originally based on domestication of roots and tubers, such as taro and yams, rice became the dominant source of food. Early agriculture in the western hemisphere spread from origins in Central America much later than in the other regions. Cropping had not reached the Great Lakes region of the USA and Canada, and the Rio de la Plata area of Argentina and Uruguay by 500 CE (Christian era).

Effects of Domestication on Plants

Variation in plant communities

While the origins of agriculture and the effects farming had on human civilizations are debated, the effects plant domestication had on plants and plant environments are rarely considered. Plant communities in natural ecosystems are marked by substantial diversity. This diversity, both between and within plant species, contributes to the survival of both the larger plant community and separate plant species. Individuals within ecosystems evolved to take advantage of this variation. For example, in natural forest communities, taller trees that only grow well in full sunlight intercept most of the sunlight. Understory plants such as ferns, mosses, and lichen, receive only the remaining light and, consequently, have evolved for shady conditions.

Diversity itself within an ecosystem helps assure survival of individual species. Besides sharing the solar radiation from above and nutrients and water from below, the variation within a plant community affords some protection against pests. Many organisms that depend on their survival by damaging plant tissue, such as viruses, bacteria, and fungi, have a limited range of hosts. For example, fungi that exist by destroying strawberry leaf tissue cannot grow on oak leaves. An insect that consumes a fern plant will not generally live on maize. Therefore, when the diversity is large, the populations of predator organisms are held in check and the chances for the survival of a host species are improved.

An example of the advantage of community diversity is a peanut-specific fungus attack on a peanut plant in a natural ecosystem. Initially, the plant is likely to suffer some damage when conditions are conducive to fungal growth. After the fungi begin to reproduce, spores are distributed through wind or splashing of rainwater and may land on neighboring plants. If these nearby plants are not peanut plants, the fungi cannot survive. The peanut plant, through selection pressures that have existed for many generations, may have partial resistance to the fungi. The plant can produce peanut seed even when infected by the fungus, since the amount of infection is not too great. Levels of the disease organism will not build up to devastating levels because nearby plants are not hosts of the fungus and cannot be attacked by it. Thus, the diversity of the plant community has a positive effect on its survival.

This concept has broader implications beyond just plant survival from disease pressures. Any condition detrimental to plant growth and reproduction will affect plant species differently. Under drought or flood conditions, some species will survive better than others. Temperature extremes may have limited effects on some species and detrimental effects on others. Competition for light and nutrients, or survival under pest problems, will differ among species. In a natural ecosystem, those species best able to survive under these stress environments will produce more seed and become more common in the community. Those plants less capable will be represented less frequently in the population. However, because of this diversity the community as a whole will remain productive.

Variation within plant species

The same principles that afford the buffering capacity against both biological and non-biological stresses in entire communities also apply to particular species. Comparisons of individual plants within a species growing in a given environment will show considerable variability for many traits. These plant traits may include size, color, and shape of plant parts such as leaves and fruit, and resistance to pests. Much of this variation comes from genetic differences among the plants for expression of these traits. The existence of variability in these traits is the result of many generations of evolution and natural selection. The diversity within a species provides the plant species with mechanisms to survive despite an adverse environment, competition from other plants, and potential destruction by viruses, bacteria, fungi, insects, and vertebrates. This diversity gives a buffer to the species in the same way the variation among species in a plant community provides protection.

Effects of domestication on variation

Understanding the process of domestication can give insights into many issues surrounding modern plant production. Importantly, a first step in plant domestication was to reduce or eliminate natural diversity. Therefore, an important natural mechanism for plant survival and reproduction was also reduced or eliminated. The major advantage of plant domestication, the ability to concentrate plants that were sources of human food, was also a major disadvantage in terms of plant survival. As people increased the numbers of desired plants on land they cultivated and removed undesirable plant species, diversity of the ecosystem decreased. The decreased plant diversity allowed pathogens and pests to concentrate and inflict damage on the crop. In addition, more homogeneous communities could be more adversely affected as a whole by non-biological stresses such as drought and temperature extremes.

Besides reducing the numbers of plant species grown in a given area, overall very few plant species were domesticated. About 125 crops from 40 families and 90 genera make up the vast majority of our food and fiber. Yet, 300 families and 3000 genera of plants are known to exist. It is not known why some plants were chosen over others. It may have been a combination of the larger edible parts of some plant species and the ease of propagating and growing these species compared with others.

As we have seen, plant domestication and agriculture spread from their areas of origin to other regions. Not only did early peoples move the ideas and practices of agriculture to other areas, but they brought with them the domesticated plant species. Transporting plant species to different regions moved the species from their centers of adaptation to regions with different growing conditions. Successful growth of these species often required modifications of the plant environment, such as amending soil components, additional irrigation, and manipulation of sowing dates.

Besides modifying the environment, humans also changed the appearance of and genetic makeup of many plant species. Although some plant traits are useful both in the wild and under domestication, many characteristics useful for survival in the wild were undesirable in food plants. Also, many plants contain toxic substances that appear to have a role in pest resistance. These toxic substances often have detrimental effects on humans as well. For example, human selection has reduced or removed alkaloids from yams, cyanogenic glucosides from cassava, steroidal alkaloids from potatoes, and resins from mango. While the plants are now more palatable to humans, they are also more susceptible to pest problems. Spines have been reduced from many crops, such as pineapple, lettuce, okra, and eggplant. These spines are thought to act as pest deterrents. Many wild ancestors of cereals and food legumes have pods that shatter. Pod splitting upon maturity to scatter seeds is an advantage to wild species, but it was a disadvantage under domestication because it makes harvests difficult.

Effects of modern agriculture

Modern agriculture requires a crop with more uniformity than that of the landraces. Hand harvesting of crops is physically difficult, uncomfortable, and monotonous. Mechanical harvesting has freed individuals to conduct other activities, but it requires a crop with uniform morphology and maturity. Efficient processing of crop plants also requires uniformity in maturity, size, shape, and color. Consumers demand uniform food products that conform to their perceptions of the appearance of a standard fruit type. The requirement adds to the lack of variability in crop varieties for many traits. For example, in the tomato, high vitamin A content is associated with an orange fruit color, yet orange tomatoes are not readily accepted by the buying public. Peanut germplasm exists with significant pest resistance. However, the material has a variegated red and white seed coat color. The desirable pest resistance can only be used in a variety after extensive genetic manipulation to change the seed coat color.

This added need for uniformity and consistency of crop plants creates an economic incentive to reduce variability. Combined with the other mechanisms to reduce diversity already discussed, there are strong pressures for genetic similarities in crop plants. Until recent times most farmers grew landraces of crop species. These landraces were made up of plants selected by farmers over many years. Landraces were usually specific to a given area, and they often varied from one farm to the next. The landraces were characterized by their large amounts of genetic variation. This variation allowed the individual crop species to retain some of the buffering capacity of natural plant communities.

Unlike these landraces made up of different genetic types, many current, improved varieties are composed of one or at most a few genetic types. Pressures to produce high yields, desirable processing traits, or meet specific marketing strategies have resulted in the use of only a few varieties of each crop species. For example, in the 1970s only four varieties each of flax, rape seed, and rye accounted for over 90% of Canadian production of these crops. In the USA, one peanut variety was grown on over 70% of the land area, three millet varieties accounted for all the production, and two pea varieties were the source of over 95% of the production. Because of concern over the limited variability of crop varieties, extensive efforts have been made to increase their diversity. Although more varieties are being grown today, the added varieties are often closely related to each other and to previous varieties. This has occurred because of the need to combine many desired traits into single varieties for market purposes. Chances of recovering all the desired traits in the process of variety development are much greater if many of the traits are already in a single variety.

Consequences of genetic uniformity

While crop uniformity provides many benefits for production, processing, and consumer acceptance, the concern over limited variability has been justified. If there is widespread use of genetically similar material, the crop will be more likely to suffer considerable damage from pest outbreaks or adverse weather conditions. Uniform production practices can also prove to be environmentally damaging. There is speculation that the collapse of the Mayan civilization was the result of overproduction of maize in highly managed areas. Intensive population pressures starting shortly after 500 CE may have required farming on lands that were not conducive to maize production. The Mayans may have terraced areas prone to erosion and added swamp lands with water drainage problems. In addition, continuous maize production is thought to have resulted in serious outbreaks of maize mosaic virus, a disease transmitted by the corn leafhopper. The food shortages may have led to the downfall of the Mayan civilization.

Irish farmers in the 1800s grew potatoes that traced back to only a few clones collected by explorers from South America in the late 1500s. Because potatoes reproduce through tubers, this asexual reproduction provided exact genetic copies of the original material on potato lands throughout Ireland. The late blight fungus (*Phytophthora infestans*) destroyed over half the potato production in 1846 because the crop was uniformly susceptible to this disease. Widespread starvation and mass emigration followed. It is estimated that Ireland's national population is not yet at the level that existed before the famine.

In the former Soviet Union, a high-yielding wheat variety called Bezostaia was developed in the 1950s. Although Bezostaia was adapted only to warmer climates, high yields and a series of moderate winters encouraged its production in the cool regions. After millions of acres were sown to this variety, a severe winter occurred in 1972, resulting in a shortfall of millions of tons of wheat.

Problems of domestic production

Natural plant populations do not usually suffer severe damage from biotic or abiotic stresses. This is true because plant resistance evolved over time, and the buffering capacity from diversity limits the specific damage. Domestication has removed the buffering capacity, and often improved varieties lack the genetic resistance present in many ancestral populations. Even when ancestral plant materials are used to introduce resistance, problems may develop. Unlike wild populations, the new varieties are uniform, allowing rapid development of pest populations that have successfully overcome the plant resistance. When this happens, it appears as though the plants have lost their resistance to diseases or insects. In actuality the resistance remains in the plant population. The small proportions of organisms in pest populations that formerly could infect plants are now at a much greater competitive advantage. This advantage results because they are the only organisms that can survive on the new types of plants. Because they are the only portions of the population that can survive, they become a much larger part of the population over time. Plants that were formerly resistant to the pest population now are susceptible because the pest population is different.

Plant Diversity and its Preservation

Centers of crop origin

Identifying the origins of domesticated plants can provide useful information and resources for maintaining and expanding genetic diversity. In addition, it provides a fascinating history of the origin of food crops. When plants became domesticated, traits and consequently genes were selected that conferred adaptation to farming rather than to survival in the wild. Selection pressure over time ultimately concentrated genes desirable from the human perspective into domesticated species. For many plant species, this process produced cultivated species very different from wild ancestors. Although ancestral plants may appear to have little to offer, there has been an increased realization that wild ancestors contain many desirable traits that are yet to be exploited. These include characteristics such as resistance to insects or diseases, variation for food quality, and improved plant architecture for mechanical harvesting.

In some cases, it is possible to identify ancestors of current crop plants with some certainty. Wheat closely resembles its wild progenitors, and it is generally agreed that its origin is in the Near East where these wild relatives are still abundant. The evolution of wheat can be traced in part by close examination of its chromosomes. Each plant species has a defined number of different chromosomes (labeled as the variable x). In wheat, the value of x is seven, i.e. there are seven different chromosomes. In the cells of most plants, chromosomes are paired so that there are two chromosomes of each type ($2x$). Species with $2x$ chromosomes are labeled as diploids. This is the case in the wild wheat species *Triticum monococcum*, or emmer wheat. A mutant occurred so that the number of chromosomes doubled ($4x$) giving rise to the species *T. turgidum* (Fig. 3.1). An apparent cross between *T. turgidum* and *T. tauschii* that was followed by a doubling of chromosomes resulted in the species *T. aestivum* ($6x$). Because *T. aestivum* has six sets of chromosomes, it is labeled as a hexaploid species. *Triticum aestivum* is the wheat species that is used in modern crop production.

Other crop plants show little resemblance to any wild species. Presumably, their wild progenitors were lost in antiquity. Maize is a case in point, since it is

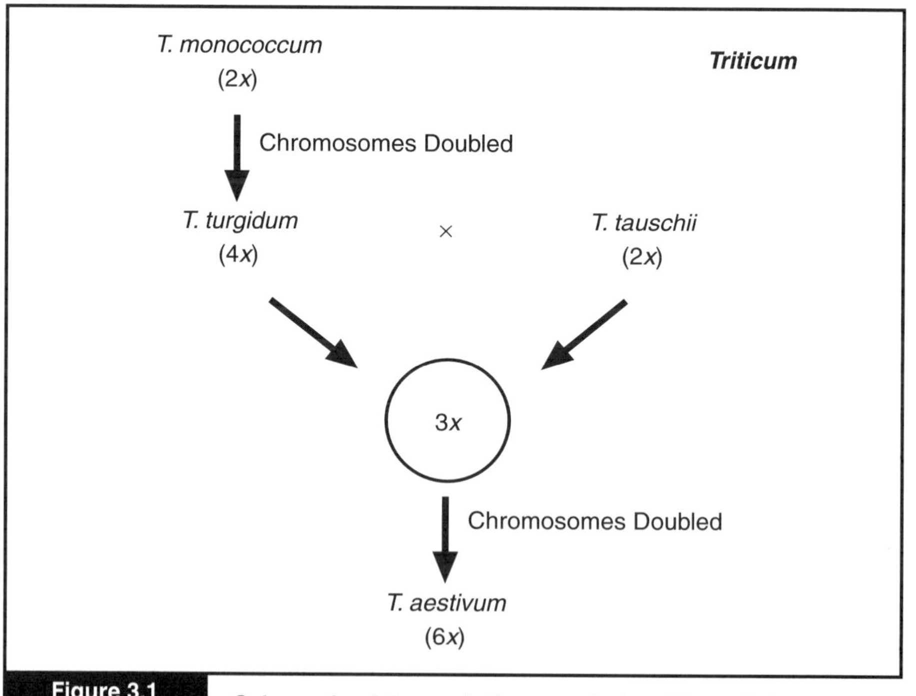

Figure 3.1 ***Schematic of the evolutionary relationship and chromosome changes among various species of wheat (*Triticum*). The value of* x *for one set of chromosomes in wheat is seven.***

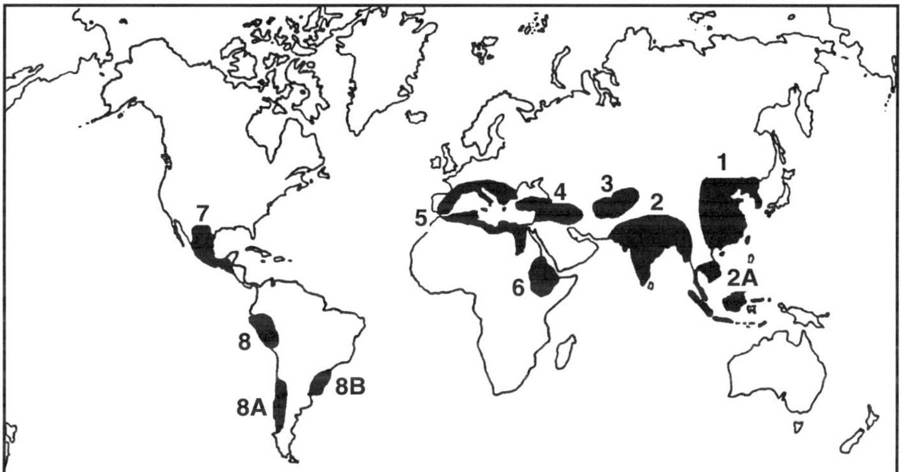

1. Chinese center: adzuki bean, millet, naked oats, sesame, soybeans.
2. Indian center: rice bean, chickpea, aboreum cotton, jute, finger millet, mungbean, rice, sugarcane, taro, yam.

2A. Indomalayan center: banana, coconut, yam, sugarcane.

3. Central Asiatic center: chickpea, flax, lentil, pea, rye, safflower, sesame, bread wheat.
4. Near Eastern center: alfalfa, barley, chickpea, flax, lentil, melon, red oats, pea, rye, sesame.
5. Mediterranean center: broad bean, cabbage, lettuce, hulled oats, durum wheat.
6. Ethiopian (formerly Abyssinian) center: barley, chickpea, flax, lentil, finger millet, pea, sesame, teff, tetraploid wheat.
7. South Mexican and Central American center: common bean, corn, upland cotton, cucurbits (gourd, squash, pumpkin), sisal, hemp.
8. South American (Peruvian–Ecuadorian–Bolivian) center: lima bean, sea-island cotton, potato, sweet potato, tobacco, tomato.

8A. Chilean center: potato.

8B. Brazilian – Paraguayan center: cacao, manioc, peanut, pineapple, rubber tree.

Figure 3.2 ***World map of centers of origin for a number of crop plants proposed by N.I. Vavilov*** *(Poehlman, 1987).*

known only as the domesticated species it is today. Many hypotheses on maize origin have been advanced. The most prevalent hypothesis is that it evolved from the wild species, teosinte, which is a weed commonly found surrounding maize patches in Mexico and Guatemala. Maize and teosinte, both diploid species with 20 chromosomes, will hybridize to produce fertile hybrids.

For many crop plants it is possible to identify areas of the world where ancestors still grow wild. These areas, called centers of origin, are determined from an understanding of the relationships between wild and cultivated species, and from an understanding of plant geography. The greatest detective in deciphering the origins of crop plants, and by inference, of agriculture, was V.I. Vavilov. Vavilov, a Russian scientist active in the early 1900s, investigated the locations where major crop species were grown. He then examined the concentrations of diversity for various plant traits such as morphology, genetics, resistance to pests, and adaptation to different climatic conditions. He also examined plant species identified as relatives of existing crop species and mapped their location. From these activities, Vavilov identified eight centers of origin. They are listed in Fig. 3.2, along with many of the plants originating from these sources. (The career of this eminent scientist ended prematurely because his ideas on genetic control of plants conflicted with strict, Marxist environmentalism dictated under Stalin.)

Some details of Vavilov's ideas have been challenged. He had limited contact with the African continent and undoubtedly under-represented crops with origins from that area. In addition, his theory provides that the greatest diversity of plant material would be found in the center of the origin rather than its margins. Others have suggested that the margins would allow more interaction between a given species and other types, thus giving greater diversity to these areas.

Despite concerns about the details, Vavilov's theory provides a working model for regional origins of modern crops. These ideas offer useful suggestions about where the greatest amount of genetic variation in ancestral crop species might be found. Most plant breeding programs around the world rely on wild plant ancestors as a source of important genes. Wild plants might provide genes for disease and insect resistance, improved nutritional quality of the food product, better plant architecture for growth and harvest, and for many other traits. To improve plants through breeding, genetic variability must exist. The current loss of genetic diversity worldwide because of population growth and human activities is a matter of serious concern. Species disappear at the rate of some 15,000 each year, causing a loss of germplasm essential to ecosystems and potentially essential genes for plant improvement. The tropical rainforests are the richest sources of genetic diversity, since they contain about 50% of Earth's estimated 5 million species.

Furthermore, the very success of plant breeding programs has caused problems. Improved cultivars often have higher yields and better pest resistance than existing varieties. The old landraces of crops that sometimes trace to primitive agriculture have been lost as farmers choose to grow the newer, higher-yielding varieties. Lost with the landraces are additional sources of genetic variability.

Plant exploration, collection and preservation

Collection and preservation of these plants are important for maintaining genetic variability. From early times, plant explorers have collected germplasm for human use. Emperors in China gathered plants for extensive medicinal gardens as early as 2800 BCE. Records exist of a collection trip in 1495 BCE sponsored by the Egyptian queen Hatshepsut to obtain trees that produced the valued resin, frankincense. Many trips during the Middle Ages have been described. Most European explorers in the sixteenth to eighteenth centuries col-

Table 3.1. International research centers with specific responsibilities for collection and preservation of crop germplasm.

Center	Location	Crops
International Maize and Wheat Improvement Center (CIMMYT)	Mexico	Maize, wheat, barley, triticale
International Rice Research Institute (IRRI)	Philippines	Rice
International Crops Research Institute for the Semi-arid Tropics (ICRISAT)	India	Sorghum, pearl millet, chickpeas, pigeon peas, peanuts
International Center for Agricultural Research in Dry Areas (ICARDA)	Syria	Wheat, barley, broad beans, lentils, cotton
International Center of Tropical Agriculture (CIAT)	Colombia	Beans, cassava, tropical forages
International Center of Tropical Agriculture (IITA)	Nigeria	Cowpea, cassava, sweet potatoes, yams
Asian Vegetable Research and Development Center (AVRDC)	Taiwan	Mungbeans, soybeans, tomatoes, chinese cabbage
International Potato Center (CIP)	Peru	Cowpea, cassava, sweet potatoes, yams

lected plants on their journeys and brought them back to Europe. Botanical gardens were developed both in Europe and in the tropical sites of European colonial possession to maintain the plants that were collected.

Most of these explorations were based on a desire either to gather plants with medicinal value or simply to obtain unique plant life. The emphasis was placed on procuring as many species as possible. Later plant explorers realized the value of diversity within species; it was the desire to gather this diversity that was the basis for Vavilov's collection trips in the 1920s and 30s. His collections led to his conception of centers of origin of crop species.

More recently, systematic collections have taken place around the world, often sponsored by governments or international research centers. Since 1974, exploration and conservation of plant genetic resources have been coordinated by the International Board for Plant Genetic Resources. This Board has two major activities: identification of species where collecting needs are most pressing, and determination of areas in the world where collecting germplasm is a priority. In recent years, controversy has surrounded the collection of materials in one country and their transfer to another country. Unlike the sale of plant commodities such as timber, minerals, and food, the seed-source country invariably receives no economic benefit from the collection of its germplasm resources. Some individuals advocate preservation at the site of origin, where wild ancestors of crop species remain in their natural state and can be controlled by the host country. Some express concerns that economic and political instability may limit access to these sites. In addition, these sites may become difficult to maintain against development pressures from expanding populations. An alternative proposed to center-of-origin preservation is continued collection of seed or other propagative material, with samples being divided to provide material for the host country's own use. Other methods of remuneration for the source country are being discussed at the international level.

Facilities exist around the world that specialize in collecting, cataloging, and storing germplasm of certain crop species. A number of international research centers have responsibilities in these areas (Table 3.1). The plant material that is collected serves as a repository for germplasm preservation and as a working collection used routinely by breeders seeking genes for improving crop plants. There is also a network of four major regional plant introduction centers throughout the USA that function in this regard. Qualified individuals receive upon request small quantities of seed for research purposes. The centers also provide medium-length storage of their accessions. Long-term storage is a safeguard against ultimate loss of the material. The US National Seed Storage Laboratory in Fort Collins, Colorado, is such a storage facility. Most seeds are stored there at temperatures below −70°C, a condition that will allow seed to remain viable for many years.

Unlike seed storage, the preservation of germplasm from crops propagated vegetatively presents a very different and more difficult problem. These species include potato, yam, sweet potato, and cassava. These root crops are the primary food source of over 500 million people for many reasons: they have yield advantages, tolerance to low fertility and drought, pest resistance, year-long season of harvests, and soil protection. Consequently, these crops are second only to cereals in meeting human carbohydrate demands, and dependence on root crops is increasing as populations continue to expand. Yet, roots are much more difficult to store than seeds on a long-term basis. Roots need to be maintained at a high water content for them to act as propagative tissue. They will not survive long-term storage either under ambient conditions or under low temperature and humidity. Storage of tissue cultures has been attempted, but there is a concern that genetic integrity of the tissue may not be maintained.

Plant Genetics

Increasing genetic diversity in modern crops is an important challenge. Genetic forces that affect variability must be understood to appreciate how diversity can be increased. The original source of variation is mutation. The variation it creates can be distributed throughout a plant population by recombination. Selection by humans or by nature will shift the amount of variation. Species migration from one location to another by natural or human forces can also change variation. Lastly, genetic drift can change the makeup of a plant population.

Mutations are changes in the genetic makeup of an organism. These changes can take place when outside sources, such as radiation or certain chemicals, cause a change in the structure of DNA. Mutations also take place when mistakes are made in the biological process of replicating genetic strands of DNA. This can occur when organisms are undergoing growth, and information from individual cells is being copied in the process of mitosis. It can also occur when organisms undergo the production of gametes for sexual reproduction, and divide half their genetic information into cells in the process of meiosis.

Although organisms have mechanisms to repair damage to DNA and the copying process by mitosis or meiosis is remarkably accurate, mutations will occur. In mitosis and meiosis, mistakes usually occur about once every 100,000–1,000,000 times a gene is copied. Usually, the mistake does not affect the organism. However, occasionally the mutation provides new variation in the expression of a trait. Because mutations are random events disturbing a process that is highly accurate, most mutations are undesirable. Under certain circumstances the mutation can produce a characteristic that has some benefit to a plant. The other forces of genetic diversity can act on this initial creation of variation to increase its frequency and spread the trait throughout a population.

Recombination is the process of repackaging genetic variation in different ways. Plants that reproduce through sexual means, usually by seed, will contribute half their genetic makeup, either as a male or female parent, to their offspring. Some plants are cross pollinated, which usually means offspring receive their genetic information from separate male and female plants. This process allows for a constant mixing of genetic information from one generation to the next and maintains high levels of variability. Other plants are self pollinated, which means that the male and female parents are the same plant, and that populations usually have less variability than corresponding cross-pollinated types. Crop plants have more self-pollinated plants than would be expected based on the frequency of self-pollinated plants in natural populations. In fact some species, such as the domesticated tomato, are self-pollinated, while most of the tomato's wild ancestors are cross-pollinated. The reason for this difference is not known. One explanation is that cross-pollination often relies on certain environmental conditions or a precise species of insect for pollination. When plants are moved from their areas of origin, the mechanisms for cross-pollination may be less efficient, or may not exist. Self-pollinated plants, that do not rely on exterior factors for seed production, may be more readily moved from their center of origin without adversely affecting seed production.

Selection is another means of affecting genetic variation in a population. This process shifts the frequency of genes in a population, often reducing the amount of variation. Those genes that code for a trait that gives an advantage to a plant will allow the plant to produce more seed and become more common in a population. Genes that code for a trait without the advantage will become much less frequent. For example, genes for more extensive root systems will allow plants containing these genes to obtain more water than other plants during drought. As a result, the extensive-rooting plants are likely to produce more seed than other plants without these genes. The genes will be passed on to following generations, and a higher percentage of plants will carry the genes for extensive root systems. As was noted earlier, similar processes took place when

Table 3.2. Number of different types of chromosomes (x), and the number of sets (ploidy) of these chromosomes in important crop species.

Common name	Chromosome number (x)	Ploidy
Alfalfa	8	$4x$
Almond	8	$2x$
Apple	17	$2x$
Apricot	8	$2x$
Barley, cultivated	7	$2x$
Cherry, sour	8	$4x$
Cherry, sweet	8	$2x$
Cotton, Asiatic	13	$2x$
Cotton, upland	13	$4x$
Oats, cultivated	7	$6x$
Orange, sweet	9	$2x$
Peach	8	$2x$
Pear	17	$2x$
Potato	12	$4x$
Strawberry, cultivated	7	$8x$
Tobacco, cultivated	12	$4x$
Wheat, bread	7	$6x$
Wheat, durum	7	$4x$

people removed characteristics from wild ancestors that were undesirable for human consumption.

The development of an entire plant population from a few individuals is called the founder principle. This principle is thought to be especially important in crop plants for several reasons. As we have noted, the spread of agriculture included the transport of seeds to new locations. These seeds were from a few plants offering little variation in the plant population in the new location.

The founder principle is also indirectly associated with another phenomenon found in crop plants. Compared with natural plant populations, domesticated species tend to have more copies of chromosomes, as illustrated in wheat in Fig. 3.1. Animals and many plants contain one copy of genetic information from the female parent and a second copy from the male parent (diploid). Crop plants, however, often contain a total of four or six copies of genetic information (polyploid) (see Table 3.2). There are several mechanisms by which polyploids may arise. One mechanism is the crossing of related species and subsequent doubling of chromosomes. A second mechanism is the development of gametes that contain a full complement of genetic information rather than the normal half. Regardless of the mechanism, development of polyploids is quite rare. It is likely the events occurred once or, at most only a few times in nature to produce the polyploids that exist today. These rare events provided little original variation when the polyploids were produced. The fact that the polyploids could have survived at all is likely to be related to their appearance. Extra genetic information in polyploid plants generally results in the production of larger plant parts. The larger plant parts have an increased chance of being noticed and saved by humans.

Countering the narrow genetic base in most polyploid species caused by the method of their creation is the genetic flexibility obtained by their extra information. Instead of having two copies of each gene that must function to produce a gene product essential for growth, polyploids have four, six, or more copies of each gene. The extra copies are not needed for the plant's survival and are free to mutate. Increased amounts of genetic information may be responsible for the considerable genetic variation usually found within polyploids.

When polyploids are originally produced, they rarely produce as many seeds as a normal diploid. If a polyploid plant can reproduce asexually, that is by cuttings rather than by seed, this difficulty is not important. This may be a partial explanation for the high frequency of asexually propagated polyploids among the most important crop plants. These species include potato, sweet potato, sugarcane, banana, yam, and strawberry.

As indicated previously, most polyploid species propagated by humans arose naturally. In some instances, researchers have produced polyploids artificially. Whether occurring naturally or produced artificially, the newly made species can only be created from plants that are closely related. Most attempts to develop economically important plants have failed. When radish and cabbage species were combined, the new plants had the top of a radish and the bottom of a cabbage. Tomato and potato were combined with the top of the potato and the bottom of the tomato. One successful creation has been the combination of rye and tetraploid wheat to form the species triticale. This cereal combines the grain yield capabilities of wheat and the greater environmental adaptability of rye.

Molecular biotechnology

Until recently, genetic variability could come about only through the forces mentioned previously. The only option of creating variability in a species was by mutation. In almost all instances this procedure was unsuccessful, since mutations are random changes in the genetic code that are almost always deleterious to the plant rather than beneficial.

Beginning in the early 1980s, a series of procedures was developed or refined that provided an entirely new way for geneticists to increase variability. Researchers are now able to isolate the genetic code responsible for expression of individual traits from virtually any organism. In principle, large quantities of the specific code (DNA) for a trait can be inserted into another organism. These procedures have important ramifications in improving our understanding of the genetic process and in applying these techniques in many areas for a better quality of life. For plant researchers studying genetic variability, these molecular genetic procedures mean the natural variation present or inducible within a species no longer limits the expression of traits.

There are several important steps in the process of incorporating DNA from another organism, a virus for example, into a plant. First, the specific DNA sequence of interest must be isolated from all the other DNA in the organism. Enzymes, called restriction endonucleases, cut DNA at the same site each time they are used. This allows the generation of specific DNA fragments that can then be isolated. A technique called the polymerase chain reaction (PCR) is used to generate a large quantity of individual fragments. Several procedures are used for incorporating DNA into plant cells. One procedure propels the DNA into meristematic regions, actively growing and differentiating regions of the plant. Another method uses the bacteria that cause crown gall tumors, *Agrobacterium tumefasciens*, to carry desired DNA into cells when they infect a plant. The desired DNA is incorporated into the plant's cells, but the bacteria itself have been engineered so that crown gall tumors do not form.

After the initial successful incorporation of foreign DNA into plants, many research groups began promising complex transformation of plants. It is now recognized that many plant traits require the transformation of many genes that have not yet been identified or cannot be moved to other species as a unit. The behavior of genes in a foreign background cannot be predicted accurately, and many desirable transformations that seemed theoretically possible have not worked in actuality. The function a gene performs in one organism may be very different from its function in another organism. In the normal host, the gene interacts with many other genes and gene products in the organism. It is also sensitive to the environment specific to the cell and to the tissue at specific stages of development. Improved understanding of many basic biological principles must occur before we can predict what a gene might do in a completely foreign background.

Molecular geneticists have always been concerned about the safety of moving DNA from one organism to another. In the early days of research in this area, scientists voluntarily curtailed experiments until the safety and ethical ramifications of placing viral genes into bacteria normally found in the human gut (*Escherichia coli*) could be determined. This may have been the first instance of scientists addressing these issues before the research was conducted. Scientists continue to be concerned about many 'what if?' questions raised as DNA is moved from one organism to another. Many early concerns about genetically engineered organisms escaping from the laboratory and disrupting health or our food chains have been unfounded. Nevertheless, strict guidelines are still in place to monitor this type of research.

Despite the complexity of the process, molecular manipulation of DNA has resulted in plants with genes from many other organisms. Such plants now contain resistance to pests or have other traits that allow food production at a lower cost to the consumer and with less impact on the environment. Research has also developed plants that produce better quality food products. In addition, some experiments have shown that it may be possible to use plants as 'factories' to provide a more efficient way of producing pharmaceuticals or other valuable chemicals. Actualization of the commercial value of these plants may rest less with the scientific feasibility than with regulatory control.

Summary

Modern agriculture has developed into a highly efficient food production system. The manipulation of both plants and their environment for maximum food production is a continuation of the process that began with domestication. Plants did not evolve to grow in fields with single or limited numbers of species. For plants to be productive in these conditions meant that the genetics of the plant and the environment had to be manipulated. It is now recognized that original ancestral plants of modern crops offer an important genetic diversity and resource for improving modern plant varieties. Methods for preserving this germplasm resource and for introducing new genetic variability in plants are important tools for increasing the productivity of plant production ecosystems.

Further Reading

Boyden, S. (1992) *Biohistory: the Interplay Between Human Society and the Biosphere – Past and Present*. UNESCO, Paris and The Parthenon Publ. Grp, Park Ridge, New Jersey.

Cook, C.M. (1991) *Genetic and Ecological Diversity: the Sport of Nature*. Chapman and Hall, London.

Evans, L.T. (1993) *Crop Evolution, Adaptation and Yield*. Cambridge University Press, Cambridge.

Ford-Lloyd, Brian and Jackson, Michael (1986) *Plant Genetic Resources: an Introduction to Their Conservation and Use*. Edward Arnold, London.

Harris, D.R. and Hillman, G.C. (1989) *Foraging and Farming: the Evolution of Plant Exploitation*. Unwin Hyman, London.

Miller, D.R. and Rossman, A.Y. (1997) Biodiversity and systematics: their application to agriculture. In: Reaka-Kudla, M.L., Wilson, D.E. and Wilson, E.O. (eds) *Biodiversity II*. Joseph Henry Press, Washington, DC.

Plucknett, Donald L., Smith, Nigel J.H., Williams, J.T. and Murthi Anishetty, N. (1987) *Gene Banks and the World's Food*. Princeton University Press, Princeton, New Jersey.

Poehlman, J.M. (1987) *Breeding Field Crops*. Van Nostrand Reinhold, New York.

Development of Agricultural Ecosystems

4

D.E. McCLOUD

Plant production ecosystems that impose the greatest control on the plant environment are commonly those associated with agriculture. Agricultural ecosystems are frequently established by eliminating the native ecosystems, tilling the soil, and controlling the invasion of desired organisms. These practices can have both long-term and long-range influences on various environmental issues. To fully understand these environmental issues it is necessary to examine the development of agricultural ecosystems.

In this chapter, the historical changes in agriculture from the earliest practices to modern crop production are examined to understand their effects on the environment. Anthropological evidence shows a humanoid presence on Earth for at least a million years. During most of this time the Earth was occupied by Paleolithic creatures. On a geological scale agriculture is a recent development and dates back only 10,000–12,000 years to the Neolithic Age.

As we go back in time, the area occupied by people on any continent, or on the Earth as a whole shrinks rapidly. In Europe, Paleolithic hunters occupied most of the area during and after the later phases of the Ice Age. The slow spread of cultivation from the east and south began only as late as 3000 BCE.

Food Gatherers

Prehistoric people obtained food by collecting plant materials and hunting animals, and therefore, they are labeled as 'food gatherers'. Food gatherers existed in a range of climates and vegetation regions, from equatorial forests to the tundra. Today, there are isolated peoples who still exist as food gatherers. These people can be found in the arid interior of Australia, the deserts and semideserts of South Africa, and the tropical rainforests of the Amazon and Congo Basins. Several food gathering communities are also scattered over the forested areas of South Asia and the adjacent islands stretching to New Guinea.

The diet of food gatherers is commonly marked by the dominance of a single resource in the native ecosystem. Some examples of this linkage between the food resource and the food gatherers are the large wild game of the Bushman territory of Africa, still more narrowly the bison of the western US plains, the salmon and marine life of Canadian coastal British Columbia, the wild reindeer of northern Asia and the North American tundra, and the sea mammals of the Arctic. Groups subsisting on vegetable staples, like the acorn of central California, are also found.

The contrasts in diet between food gathering peoples may be extreme. The Eskimo and the Plains hunters lived almost entirely on meat, and peoples in coastal British Columbia and eastern Siberia lived largely on fish. However, it should be noted that such specialization or limitation is never complete. The once postulated pure gleaner who exists solely on gathered fruits, insects, and the smallest animals caught by hand remains a hypothesis unconfirmed by ethnological investigations. Eskimos, who exhibit the closest approach to a purely hunting and fishing subsistence, regularly gather berries during their summer travels.

The geographical distribution of food gatherers throughout the world before agricultural development was not uniform, but consisted of pockets of population where resources allowed mankind's development. The population that could concentrate in a small area was small even where the resources were great – usually less than one person per square kilometer (100 ha per person). Food gathering activities required very large amounts of land, and the impact on the natural ecosystem was small.

Food gatherers had as their most important characteristic the lack of any attempt to increase the relative occurrence of specific plants or animals. This situation placed them very much at the mercy of the native ecosystem, and food consumption tended to vary greatly from place-to-place, day-to-day, season-to-season, and year-to-year. Periods of abundance alternated with periods of severe shortage. Often life itself was at stake. Food gathering usually offered no more than a marginal existence and exposed its peoples to frequent hardships, inadequate nourishment, ill health, periodic famines, and starvation. The natural ecosystem usually supported only very few people per land unit, and even then, often did so only by condemning the people to subsistence living. However, the natural balances of the ecosystem in which food gatherers were integrated, imposed a check on human consumption and restricted opportunities for population increase.

Anthropologists recognize the Mesolithic Age as the transition stage between food gathering and agriculture. Mesolithic people used more refined stone tools and began to manage and use some plants, but they only developed a pre-agricultural society. The Mesolithic Age saw the first settled society and both domestic plants and animals were developed. Highly polished stone tools were produced, weaving began, and pottery was developed. However, this society could usually afford only one member not involved in the accumulation of food: the medicine man who appeased the gods.

Agricultural Ecosystems

The development of agricultural ecosystems can be classified into four broad categories: shifting cultivation; sedentary subsistence farming; livestock farming; and industrialized agriculture. Industrialized agriculture can be further subdivided into four levels based on the level of fertilizer use and the resultant crop yield.

Shifting cultivation

Agriculture dates back only about 10,000–12,000 years to the Neolithic Age. Agriculture began in the Tigris–Euphrates, the Indus, and the Nile river valleys. Practically all of our current crop plants, and domestic animals, were developed during prehistoric agricultural times. The development of agriculture followed two major patterns. One system was restricted to areas of highly fertile soils, usually the river alluvium (fluvisols) or productive terrace soils (andosols). Both situations took advantage of a fertile soil that was periodically enriched by the drift of water borne or wind borne materials. The second system was one of shifting cultivation, which is also known as slash-and-burn cultivation, milpa, ladang, or chitemene. This system takes advantage of the large nutrient supply available from burning the forests or grass vegetation of the natural ecosystem. Currently, over 200 million people thinly scattered over 36 million km^2 of the tropics (about six persons km^{-2}) obtain the bulk of their food by shifting cultivation systems.

A remarkable feature of shifting cultivation is its universality. Primitive people evolving from the status of simple food gatherers, in regions widely separated, responded to the challenge of food production in the same way. In the tropics of Africa, America, Oceania, and Southeast Asia, the system was similar. These people cut and burned the forest, planted their crops with a simple digging stick, and after taking one or two harvests, they abandoned their plot to the invading forest. Each year they moved to a new patch of land to be cleared for cropping, therefore the term shifting cultivation. Later, some developed a hoe with which to clear tenacious grass roots and cultivate the savannas. In this system too, they abandoned their plots after a few years of cropping.

Although shifting cultivation is often regarded as a primitive agricultural system of the tropics, its practice was not confined to the tropics. Nor were the peoples who employed shifting cultivation necessarily primitive in technology or in culture. Shifting cultivation persisted in areas of Europe during the Middle Ages. The manorial two-crop rotation system, a form of intensive shifting cultivation, was normal in the fifth century and had not altogether disappeared at the close of the Middle Ages. The English settlers in Virginia in the seventeenth century adopted shifting cultivation from the Native Americans, since they found it was the best way to deal with the abundant forests. There really was no other system for maintaining fertility.

Shifting cultivation nearly always involves clearing by burning. It is frequently called slash-and-burn agriculture though the term swidden farming is now used by some anthropologists after the Old English dialect word, swidden, meaning a burned clearing. Shifting cultivation on forest lands supports a population of only about 7.7 people km^{-2} (13 ha per person).

When the forest is cleared and the debris burned, all the nutrient elements, except nitrogen and sulfur, are deposited on the surface as ash. Subsequent rains wash the nutrients into the soil. Nitrogen, sulfur, and carbon in the burned material are lost as gases. Potassium is present in excessive quantities. Much of the litter layer is also destroyed by burning. The nutrients in the roots are also released slowly into the soil. The total remnant material available after burning of tropical forests in shifting cultivation systems is about 1000 kg ha^{-1}. Phosphorus is commonly the most limiting element and is about 70 kg ha^{-1} of the total.

It is important to recognize the need for shifting cultivation of so many farmers in so many countries. The effort to clear trees is a large one and no one would do so unless there is a necessity. Further, having made a clearing, any sensible farmer will use it for as long as possible. Two factors are mainly responsible for shifting cultivation. First, invasion by weed species becomes troublesome, and second, nutrients are exhausted in the old cleared land. Fertile soils are not to be found everywhere, and on infertile soils, shifting cultivation is the system that makes nutrients available.

Shifting cultivation is the beginning of human alteration of the natural ecosystem because land is cleared and burning of vegetation is practiced. With low population pressures on land, however, little permanent change is effected. The difficulty is that increasing population pressures bring shorter rotation cycles and eventually a deterioration of the natural ecosystem.

Sedentary subsistence farming

The transition from shifting cultivation to sedentary subsistence farming did not result in increased yields per hectare; basic cereal yields remained at about 1 t ha^{-1}. The transition did, however, result in more intensive land use. By the Middle Ages, Western European agriculture had settled to using the better land continuously in a two-crop rotation. Under this agricultural ecosystem, the carrying capacity of the land was low.

The low carrying capacity of sedentary subsistence farming in the early Middle Ages is illustrated in the records of the monastery of St Symphorien in Antver, France. Fifteen monks and some servants in the monastery just managed to live on the excess yield produced by 100 farm families. These family farms produced only about 1 t of grain ha^{-1}, and about 200 kg of this yield was required for seed the next year. Roughly half the net yield of 800 kg ha^{-1} was needed to feed the animals that provided the muscle power for the farms, and to produce beer. Beer was needed because the quality of the water was poor and the meat was salty. The remaining 400 kg ha^{-1} of grain yield was little more than enough to feed the person who did the work.

The low yields in Western Europe resulted not so much from an adverse climate, but from a lack of plant nutrients. Expressed in traditional units of N + P_2O_5 + K_2O, only 25 kg ha^{-1} of nutrients were available each year for the growth of the crop. In sedentary subsistence farming, the main source of additional nutrients was manure. The small price differential between meat and grain, and the large meat consumption in the Middle Ages, showed that herds of grazing animals were maintained to a certain extent for their manure. Meat was more or less a by-product.

Higher yields were harvested when the potato was introduced into Western Europe. The higher yields resulted not so much from a greater production of plant material, but from a more advantageous distribution between the various parts of the plant. In potato, about 80% of the plant material is in the tuber and only 20% is in leaves and stems. Thus, the number of persons who could live off 1 ha of land was almost twice as large with potatoes as with cereals. The larger yield of potato, more than its taste, explains the rapid adoption of the potato in Western Europe.

Sedentary subsistence farming also causes more permanent impacts on both natural and agricultural ecosystems than the more primitive systems. Forests are cleared, grasslands cultivated, and nutrient transfer results around villages. This system commonly produced a gradient of fertility with higher fertility land developing closer to the villages, an effect that is readily observable in many countries even today. The pre-colonial agriculture in the subhumid highlands of central Mexico resulted in severe soil erosion about 3500 years ago. Others found similar impacts for the classic Mayan cities in the lowland rainforests of Guatemala and for the indigenous cultivators in New Guinea. These old 'traditional' forms of agriculture were commonly associated with severe environmental degradation.

Livestock farming

The first widespread yield breakthrough for crops came in the late 1600s. It was hailed as 'The New Agriculture' and caused a revolution in agriculture. Cereal yields were doubled to 2 t ha^{-1} (Fig. 4.1). This agricultural ecosystem was based on the production of cereals and other food crops, and these were rotated with clovers and grasses for feeding livestock. All the urine and manure from the animals was carefully saved for the cropland, hence livestock farming. Refuse and night soil from the towns was also collected and applied to the fields of grain crops. The total fertilizer equivalent for the livestock farming system was estimated at 50 kg ha^{-1}, and 2 t ha^{-1} of crop yield was produced.

These new, doubled yields with livestock farming were produced in Western Europe from the early 1700s until the late 1800s (Fig. 4.1). This agricultural ecosystem was used in the eastern USA until the beginning of the twentieth century. Livestock farming required more land than the sedentary subsistence ecosystem, although much more of the available land can be cultivated than with the sedentary subsistence system. This is true because manure

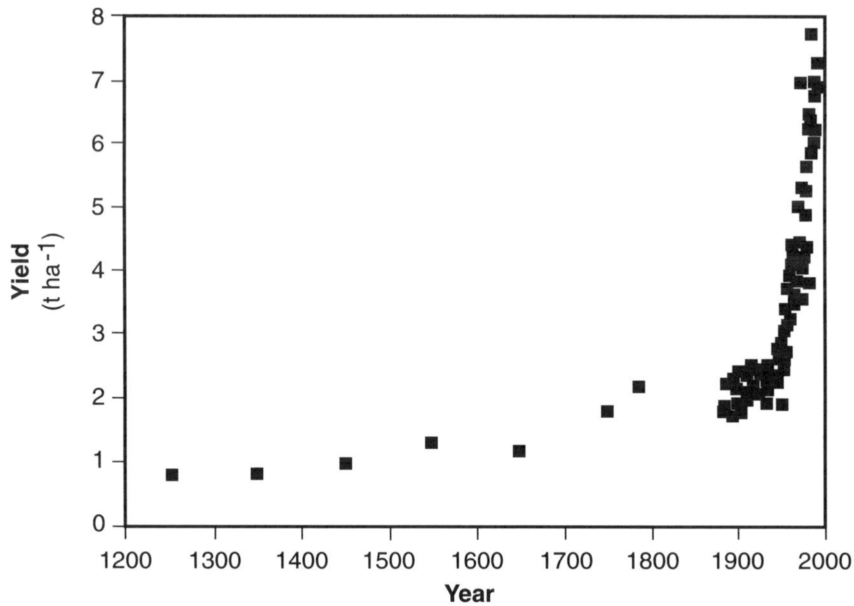

Figure 4.1 ***Wheat yields in the United Kingdom from about 1250 to present.***

can be applied to lower fertility soils and then they can be used for cereal production.

The livestock farming ecosystem had a large influence on the agricultural environment. Currently in Africa, Nigerian herdsmen are paid to herd their cattle during the dry season on to croplands to convert crop residues into manure. In India, the land around the old established villages varies in quality according to the distance from the village. Lands close to villages are known as the 'Adhan land' and they have benefited from perhaps 2000 years of casual disposal of wastes and excrement. This land is superior in both soil texture and in productivity to the more distant lands from which crops have been continually harvested and taken to the village. In Britain, there is an old proverb that says 'the nearer to the church, the better the land'. The chief reason for the superiority of the land is a long history of fertility transfer.

Industrialized agriculture

The next breakthrough for increasing agricultural yields came with the discovery and use of fertilizers. Liebig, a German chemist, provided conclusive evidence

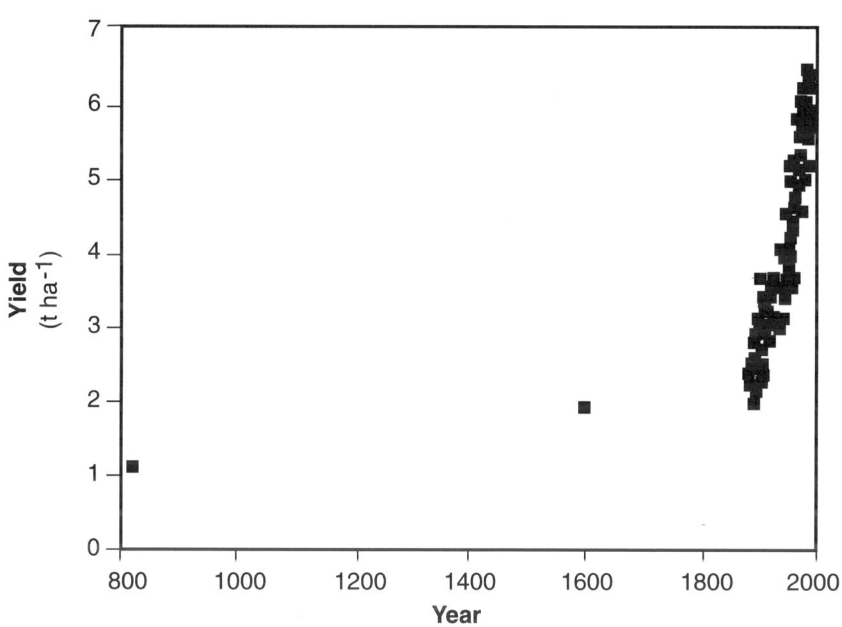

Figure 4.2 ***Rice yields in Japan from about 800 to present.***

that plants need only water and minerals from the soil for growth. He erroneously assumed, however, that nitrogen, like carbon dioxide, was obtained from the air and was never limiting. He considered that fertilization with nitrogen-containing nutrients was unnecessary. Liebig's view was opposed by Lawes and Gilbert, founders of the Rothamsted Agricultural Experiment Station in England, who showed that mineral nitrogen was essential for the growth of cereal grains. Also, Wolff in 1856 convincingly disproved Liebig's theory on nitrogen. Lawes and Gilbert pioneered the development and use of mineral fertilizers. Through fertilizer additions, the limitation of 50 kg ha^{-1} of plant nutrients was removed and yield could be increased several-fold (Fig. 4.1).

The industrialized agriculture ecosystem allowed the application of missing nutrients for specific soil situations and cultivation of land that would otherwise be unusable. Today, there is fertility transfer on a global scale, e.g. Florida marine-deposited phosphate is applied to India's wheat lands.

Nutrients and Crop Yield

For nearly 10,000 years, yield of cereal crops such as wheat remained stable at about 1 t ha^{-1}, or nearly one-half cup of grain per square yard. In Europe from

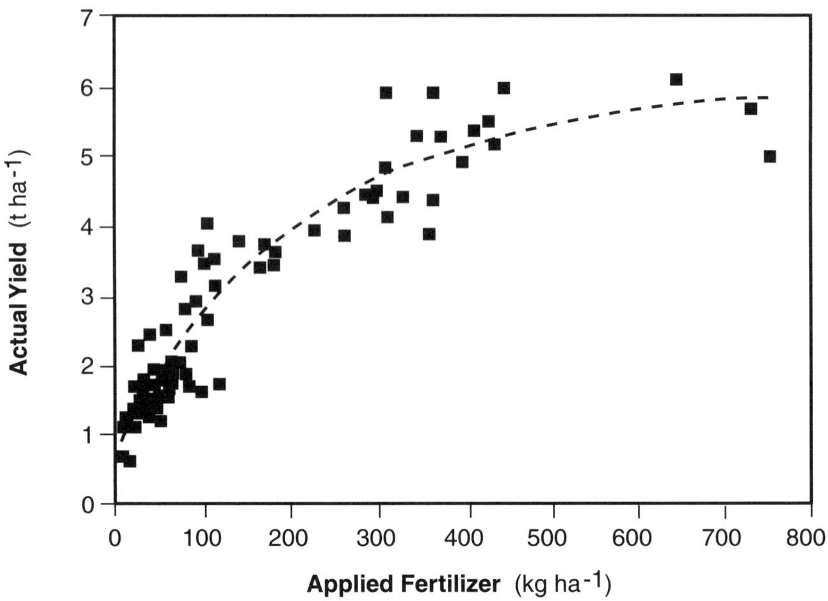

Figure 4.3 ***Annual grain yields for 18 of the most populous countries at 5-year intervals from 1973 to 1988 plotted against amount of applied fertilizer.***

the 1200s to the mid-1600s, 1 t ha^{-1} was the unvarying agricultural yield (Fig. 4.1). Increase in yield to 2 t ha^{-1} from the 1600s to the mid-1700s was a result of a shift in the agriculture ecosystem. The two-crop manorial rotation system was replaced with the four-crop rotation system of livestock farming. From the mid-1700s until the mid-1940s, 2 t ha^{-1} was the regular yield in industrialized countries. In the UK, a spectacular increase in wheat yields began in the mid-1940s. This yield revolution has been essentially a fertility revolution, with increasing levels of nitrogen fertilizer. The 1984 wheat yield of 7.72 t ha^{-1} was an all-time record yield for the UK.

Rice yields around 800 are estimated at 1 t ha^{-1} (Fig. 4.2). By 1600 rice yields were nearly at 2 t ha^{-1} because of improved practices. These improvements included the use of nitrogen fixation by the blue-green algae and of additional night soil. For the next 300 years until the early 1900s, rice yields fluctuated around 2 t ha^{-1}. Then a spectacular yield increase began, and these high yields continue to the present. The 1978 rice yield of 6.42 t ha^{-1} was an all-time record yield for Japan.

The development of the industrialized agriculture ecosystem brought vast changes in nutrient availability for application in agricultural ecosystems. Beginning in the early 1950s, increasing amounts of fertilizer were used in most developed counties to produce ever-increasing crop yields. The increase in fertilizer worldwide has been especially rapid over the past 20 years. Many developing countries have increased their fertilizer usage with a corresponding increase in crop yields.

The close correlation between fertilizer application and grain yield at the country level is readily visualized by plotting these data against each other. For 18 of the most populous countries, grain yields are plotted against fertilizer use in 1973, 1978, 1983, and 1988 (Fig. 4.3). These countries included The Netherlands which had the highest fertilizer use, and Malawi which was among the lowest fertilizer users. The increase in grain yields was very closely linked to the level of fertilizer application, and the relationship was described by an asymptotic exponential equation.

The data at zero nitrogen shows an intercept at 1 t ha^{-1}, which is the cereal yield generally obtained for shifting cultivation and sedentary subsistence cultivation. Livestock agriculture, which provided 50 kg ha^{-1} of fertilizer, produced yields up to 2 t ha^{-1}. Industrialized agriculture can be divided into four levels of fertilizer application: low fertilizer, 50–199 kg ha^{-1} (at 50 kg ha^{-1} fertilizer, yield is 2 t ha^{-1}); medium fertilizer, 200–399 kg ha^{-1} (at the 200 kg ha^{-1} level yield is 4 t ha^{-1}); high fertilizer, 400–599 kg ha^{-1} (at 400 kg ha^{-1}, a yield of 5.2 t ha^{-1} is produced); and very high fertilizer, 600–800 kg ha^{-1} (at 600 kg ha^{-1} yield is 5.8 t ha^{-1}).

In the low fertilizer range (50–199 kg ha^{-1}), yields increased 13.3 kg per kg fertilizer. The medium fertilizer range (200–399 kg ha^{-1}) produced yield increases of less than half as much – 6 kg yield per kg fertilizer. The high range (400–599 kg ha^{-1}) again produced less than half as much yield increase: 3 kg yield per kg fertilizer. This is an expression of the principle of diminishing returns in the yield–fertilizer response.

There is a growing awareness of the adverse ecological consequences that have sometimes been attached to expanding application of chemical fertilizers. There is also a limit on the availability of fossil fuels to support fertilizer production. Consequently, there has been heightened interest in alternate means of providing nutrients to crops. The possibility of making greater use of organic fertilizers – animal, human, and plant wastes – is receiving increasing attention. Sludge from human sewage treatment plants is already being used for agricultural purposes in some developed countries. The Chinese have long assiduously returned organic wastes to their fields with beneficial results.

Unfortunately, there are economic obstacles to widespread use of organic fertilizers. The nutrient content of manures and vegetable matter for composting is quite low. Massive volumes must be transported and often treated to provide the needed amounts of nutrients for crops. Health hazards may arise if the waste is not properly handled. Nevertheless, the potential usefulness of organic fertilizers deserves greater exploration, especially in the developing countries where labor is abundant and cheap.

In Southeast Asia and South Asia, the greatest obstacle to the use of animal manures for fertilizers is the critical need to use dried dung for cooking fuel. Efficient use of available organic fertilizers cannot substitute for chemical nutrients, as China's soaring need for chemical fertilizers shows. Recycling wastes

Table 4.1. Major crops, area harvested, production, principal countries of production and highest average country yield.

Crop	Area harvested 1995 (million ha)	Production at 10% moisture 1995 (million Mt)	Principal countries in 1995	All-time Highest country yield (t ha^{-1})	All-time Country, year
Rice	148	522	China, India, Indonesia	9.0	Australia '95
Wheat	227	641	China, USA, Russian Fed.	8.9	Netherlands '95
Maize	136	515	USA, China, Brazil	9.4	New Zealand '95
Barley	99	143	Russian Fed., Canada, USA	6.1	Netherlands '95
Soybean	62	126	USA, Brazil, China	4.6	Greece '90
Sugarcane	18	115	Brazil, India, Cuba	120	Peru '94
Cassava	16	72	Brazil, Nigeria, Thailand	24	India '95
Potato	18	65	China, Russian Fed., Poland	50	Belgium-Lux '93
Sorghum	41	57	USA, Nigeria, China	6.4	Italy '89
Sugarbeet	9	47	France, Ukraine, Germany	72	France '93
Cottonseed	34	39	China, USA, India	4.6	Israel '93
Sweet potato	9	30	China, Vietnam, Indonesia	9.5	Indonesia '95
Oats	18	29	Russian Fed., Canada, USA	6.4	Ireland '95
Peanut	22	28	India, China, USA	7.0	Israel '90
Millet	37	26	India, Nigeria, China	3.4	Spain '94
Rye	10	24	Poland, Germany, Russian Fed.	7.5	Netherlands '95
Total	904	2479			

Source: FAO (1995).

can, however, fulfill some limited portion of the nutrient needs and also have beneficial side-effects in native ecosystems.

Industrialized Agriculture Ecosystems

Industrialized agriculture has focused on only 16 plant species to provide most of the world's food. On a dry matter or caloric basis these 16 crops account for 92% of the food consumed by humans. Of course the other 8% of food contributes valuable vitamins and adds variety to otherwise monotonous diets. Additionally, large quantities of cereal grains and oilseed crops are used for livestock feed. The livestock contributes to the food of humans in the form of meat, milk, and eggs. Of the 16 major crops, three – wheat, rice, and maize – account for nearly two-thirds of the total production. These three comprise 60% of the area sown to the 16 crops. Data for these 16 crops are presented in Table 4.1.

Seven of the 16 major crops, wheat, rice, maize, barley, sorghum, oats and rye, are categorized as cereals. Soybean and peanut are legumes that have the unique ability to 'manufacture' the nitrogen required through a symbiotic relationship with bacteria. These legume species have high protein and oil contents. The two sugar crops are sugarcane and sugarbeets. Potato, sweet potato, and cassava are root or tuber crops. Cottonseed, a by-product, ranks 13th in terms of dry matter production; the oil is used for human consumption and the meal as livestock feed.

As expected, China with a population of over 1 billion is one of the top three countries in the production of nine of the 16 crops. The former USSR and the USA are next – each is among the top three in the production of four of the 16 crops. India with a population of 853 million, appears four times among the top three countries in production of the 16 crops.

The highest average yield for each of the 16 crops and the country and the year in which the record was set are shown in Table 4.1. Sugarcane leads with a record dry matter of 120 t ha^{-1} obtained in 1994 by Peru. Sugarcane has a photosynthetically efficient C_4 metabolism, and the yield is stored as simple sugars rather than complex carbohydrates, proteins, or oils that are less efficient to manufacture. Furthermore, much high yielding sugarcane needs 18 months to 2 years to mature. Sugarbeet is next with 72 t ha^{-1} from France in 1990. Sugarbeet, while a less efficient C_3 plant has the same advantages as sugarcane requiring more than a year to mature and accumulating its yield as simple sugar. For the grain crops the maize yield of 9.4 t ha^{-1} was an all-time record yield for this crop set by New Zealand in 1995.

Another feature of industrialized agriculture is the large increases in the productivity of each farm worker. Prior to the industrial age of agriculture, grain was harvested with a 'cradle'. The cradle was a scythe to which several lightweight sticks were attached so that the grain could be laid in windrows as it was cut. A second person followed and using a handful of cut plants tied the grain into bundles that were set in shocks to dry before threshing. One person using this system could 'cradle' only about 0.25 ha in a day. The reaper invented by a Virginia planter, Cyrus McCormick, began the industrial revolution in grain harvesting. With a reaper, one person and two horses could cut 1 ha of wheat in a day. A person with team of horses, working from daylight to dark, could harvest 2.5 t of maize from 1 ha in a day. Contrast that with the modern self-propelled harvester with a 12 meter width of cut (swath) which can harvest 100 hectares per day! However, large economic expenditures are required in carrying out industrialized agriculture. For example, the price of the enormous, air-conditioned maize/soybean harvester is about US $150,000.

In central Illinois, a typical maize/soybean grain farm with 350 ha of cropland might grow 180 ha of maize and 170 ha of soybeans. Fertilizer applications of 100 kg ha^{-1} and maize yields of 11 t ha^{-1} are ordinarily obtained by central Illinois farmers. On the other hand, soybeans are legumes and require little or no fertilizer nitrogen, however they yield only 3.3 t ha^{-1}. The much lower yield results from the high protein and oil content of the seeds of soybean compared with maize.

An example of small scale, modern industrialized agriculture is found in Japan. The fields are usually about 10 ha in size. In the Hokkaido Prefecture, a considerable area of rice is grown farther north than any other region of the world. Since the growing season is too short for normal paddy rice, plants must be seeded in a glasshouse nursery. The rice plants are transplanted to the field after the weather becomes warm enough for rice to grow. Harvesting is accomplished with a tiny mechanical harvester. Since 1975, Hokkaido average yields of rice have been above 5 t ha^{-1} and typical yields for individual farmers are much higher.

Summary

In the early stage of human development, food was obtained by food gatherers. Food gatherers were integrated into the natural ecosystems and abuses to the ecosystem resulted in human starvation, disease, and death. The development of agricultural ecosystems meant the destruction of the native ecosystem to improve nutrient availability and to deter organisms that competed with or attacked the crop plants. The shifting cultivation ecosystem supported only a low population density and was commonly associated with serious environmental degradations. Sedentary subsistence farming allowed an increase in crop yields by employing management techniques to concentrate nutrients on croplands. Simple crop rotations and the application of manure were practiced in this system. Livestock farming followed, which is characterized by a further concentration of nutrients, especially by grazing animals. The inclusion of nitrogen fixing legumes into rotations and the application of manure from grazing animals was important in this ecosystem. Finally, the industrialized agriculture ecosystem resulted in many-fold increases because of the application to croplands of chemical fertilizers.

The productivity of agricultural ecosystems at any level is very closely correlated with the availability of nutrients in the soil to support plant growth. Consequently, each of the successive stages of agricultural ecosystems is associated with techniques to provide crops with increasing amounts of nutrients. In industrialized agriculture, cereal yields are now very closely linked to the amount of applied fertilizer. The yield response to fertilizer, however, is not linear and the most advanced countries have approached saturation in increasing yield to increased fertilizer applications. Industrialized agriculture ecosystems rely on only a few plant species. Only 16 plant species account for 92% of the food consumed by humans. These plant species are grown as crops requiring large investments in machinery and fossil fuel.

Further Reading

Altieri, M.A. (1995) *Agroecology: the Science of Sustainable Agriculture*. Westview Press, Boulder, Colorado.

deWit, C.T. (1972) Food production: past, present and future. *Stickstof* 15, 68–80.

deWit, C.T. (1974) Early theoretical concepts in soil fertility. *Netherlands Journal of Agricultural Science* 22, 314–324.

Harlan, J.R. (1995) *The Living Fields: Our Agricultural Heritage*. Cambridge University Press, London.
Hutchinson, Joseph (1972) *Farming and Food Supply*. Cambridge University Press, Cambridge. 146 pp.
McCloud, D.E. (1979) *Man's Food Crop Resources*. University of Florida, Institute of Food and Agricultural Sciences.
McCloud, D.E. (1989) *Climate and Food Security: Assessing Climatic Variability from Long-term Crop Yield Trends*. International Rice Research Institute, pp. 201–217.
O'Hara, Sarah L. *et al.* (1993) Accelerated soil erosion around a Mexican highland lake caused by prehispanic agriculture. *Nature* 362.
Smith, B.D. (1995) *The Emergence of Agriculture*. W.H. Freeman Press, New York.

Environmental Limits to Plant Production

5

T.R. SINCLAIR AND F.P. GARDNER

Plant production ecosystems are designed and managed to channel effectively the flow of mass and energy into primary production and into a myriad of food chains. The primary objective is harvestable mass, but nutritional value and other quality factors may also be important. Both agricultural and natural ecosystems collect solar energy over extended periods and store it as chemical energy. The chemical energy is stored in carbohydrates, proteins, and lipids, which are about 95% of the total plant dry mass. Accumulated plant mass can be no greater than the amount of energy intercepted by the plants and used to form chemical energy. Whether dealing with a rice crop or an evergreen forest, the same basic relationships of solar energy interception and use define the ultimate limit on the amount of accumulated plant dry matter.

The interception of solar energy and its use to form harvestable plant mass in natural and plant production ecosystems can be described by four processes.

1. Daily interception of solar energy that is dependent on the area of leaves available to intercept the light.
2. Efficiency of the plants in using the solar energy to produce plant materials.
3. Summation of the total amount of solar energy intercepted during the life of the plants.
4. Allocation of the plant materials to plant parts important to the ecosystem. Of course, in plant production ecosystems the important plant parts are those harvested for the benefit of humans.

Each of these four processes can have a tremendous influence on the eventual productivity of a plant production ecosystem. Inadequate performance in any of the four processes can result in substantial losses.

Interception of Solar Radiation

The wavelengths of solar energy, or more generally solar radiation, that are critical in plant production are those that have sufficient energy when absorbed by pigments to cause changes in chemical energy levels. This portion of the spectrum is commonly called the photosynthetically active radiation, and corresponds roughly with the portion of the spectrum visible to the human eye. (A complete description of the solar energy spectrum important in plant production is presented in Chapter 8.) In this discussion of the limits to plant production, two processes are particularly important. These are the interception of the incident light rays, usually by leaves, and the absorption of the intercepted light by plant pigments.

The interception of solar radiation by plants is done by displaying various surfaces on which light might fall before it reaches the soil. Nearly all plant canopies are readily characterized by the presence of leaves to intercept solar radiation. The fraction of the solar radiation reaching the Earth's surface intercepted by a leaf canopy is dependent on the extent of the leaf surface area. In turn, canopy leaf area depends on the number and size of leaves, both of which are influenced by environment and plant genetics. Such factors as temperature, drought stress, and nutritional status of the plant can greatly affect leaf size. The genetic blueprint for the development of the leaves is quite important in the distinguishing appearance of oak leaves and maize leaves.

In estimating the interception of solar radiation, the important parameter is the leaf area index (LAI). LAI is a ratio of the leaf surface area (one side) per unit ground surface area. For many crops in high-yielding agricultural ecosystems, the LAI is roughly 4–6, i.e., 4–6 m^2 leaf area per m^2 of ground area. In forests and natural ecosystems the LAI can be much higher with the maximum reported LAI for a Scots pine forest of 18. As a practical matter, LAI values seldom exceed 10 since lower leaves usually senesce in the very low light levels at the bottom of the canopy when new leaves are added above them.

The fraction of the daily solar radiation intercepted by a leaf canopy (F) is generally well described as a function of LAI by a simple exponential equation.

$$F = 1 - \exp(-K \times \mathrm{LAI}) \tag{5.1}$$

where K = equal extinction coefficient (approximately 0.7).

The value of K is dependent on several factors involved in the geometry of light interception. These factors include the horizontal and vertical leaf distribution, angle of the leaf surfaces, angle of the sun, and the amount of diffuse radiation. Nevertheless, for many situations the value of K that represents the daily interception of photosynthetically active radiation is stable at a value of about 0.7.

Equation 5.1 shows that the LAI is critically important in determining the fraction of solar radiation interception. The value of F increases asymptotically with increases in LAI. About half the radiation is intercepted with an LAI of only 1.0, and 90% of the radiation is intercepted at an LAI of 3.3. An LAI of 4.3 is needed to intercept 95% of the solar radiation. Four cultural factors promote the achievement of high LAI: (i) adequate supply of resources for plant development, especially water and nitrogen to promote rapid leaf expansion; (ii) a vigorous-growing genotype (rapid leaf expansion rate); (iii) adequate plant den-

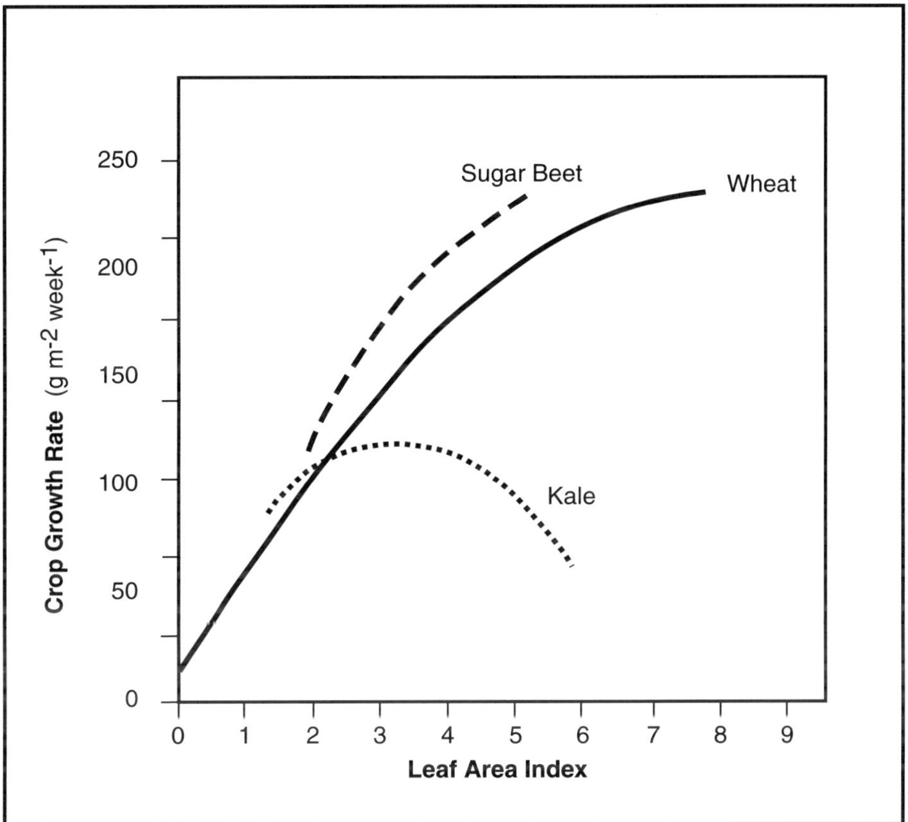

Figure 5.1 ***Crop growth rate on a weekly basis plotted against leaf area index.***

sity; and (iv) a planting pattern with optimum spatial arrangement (equidistant is superior).

Not surprisingly, the growth rate of many plant systems increases with increasing leaf area. Because *F* increases asymptotically with LAI, an asymptotic increase in crop growth rate with increasing LAI has been observed. Figure 5.1 is a plot of experimental results for wheat and sugarbeet showing the saturating growth rate as LAI increases.

Determination of the actual amount of solar energy intercepted (*I*) is based on a multiplication of *F* by the amount of radiation incident to the plant canopy (*Io*). That is,

$$I = F \times Io \tag{5.2}$$

Therefore, the actual intercepted radiation is dependent on how much incident radiation is available in a particular environment. Environments where the daily amount of incident radiation is great have a greater potential for high levels of intercepted radiation. Desert regions with clear, cloudless skies typically have large values of *Io*.

Radiation Use Efficiency

The intercepted solar energy is absorbed in leaves and used to produce chemical energy in the plant. The pigment that absorbs the solar radiation is chlorophyll, which gives leaves their familiar green coloring. Chlorophyll appears green because it absorbs the green wavelengths in the solar spectrum slightly less than the other photosynthetically active wavelengths. The energy absorbed by chlorophyll causes the pigment to exist briefly in an 'excited' state that represents chemical energy. This energy is rapidly transferred to the energy-rich intermediate metabolites, adenosine triphosphate (ATP) and reduced nicotinamide adenine dinucleotide phosphate (NADPH). These intermediate metabolites, in turn, provide the chemical energy required for assimilating atmospheric carbon dioxide (CO_2) into simple organic molecules. Additional chemical energy is provided to synthesize these short-chain molecules into all the various complex biochemical constituents of plants.

The initial fixation of CO_2 into organic molecules is done by one of the most abundant proteins on Earth: ribulose-1,5-bisphosphate carboxylase/oxygenase (RUBISCO). The great abundance of RUBISCO is explained in part by the fact that RUBISCO has only a modest attraction for CO_2 molecules.

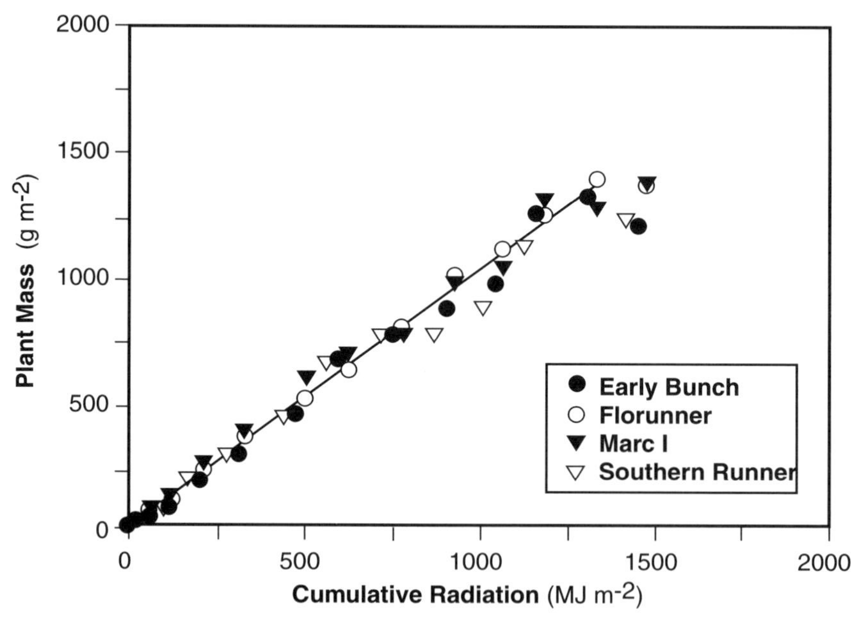

Figure 5.2 ***Accumulated plant mass for four varieties of peanut plotted against cumulative intercepted solar radiation. The slope of this plot is equal to a radiation use efficiency of 1.0 g MJ^{-1}*** *(Bennett* et al., *1993).*

RUBISCO is needed in quantity to capture CO_2 at a reasonable rate. Once a CO_2 molecule is captured by RUBISCO, the carbon atom is quickly passed along to other molecules in the Calvin cycle using the energy supplied by ATP and NADPH. Eventually, simple three-carbon sugars (C_3) are generated and used as the building blocks for all other compounds in plants. Those plant species that use only this process to assimilate CO_2 are identified as C_3 species. Most species used in plant production systems are C_3 species.

A group of plants has evolved in which the preliminary fixation of CO_2 is done by phosphoenolpyruvate carboxylase (PEP). PEP has a high affinity for CO_2 and therefore, can capture CO_2 at a high rate with a lower amount of protein. The initial fixation of CO_2 by PEP produces four-carbon organic acids. Hence, the species that have the preliminary PEP fixation process are identified as C_4 species. The C_4 organic acids are transported into specialized cells in the leaves of these species where the CO_2 is released to RUBISCO and Calvin cycle metabolism. Important C_4 species in plant production include such warm-season grasses as maize, sorghum, and sugarcane.

In either the C_3 or C_4 species the amount of CO_2 assimilated is directly dependent on the amount of solar energy that has been stored in ATP and NADPH to fuel the Calvin cycle. Consequently, in plant production, the accumulation of dry matter is directly related to the amount of absorbed solar energy. As the amount of absorbed photosynthetically active radiation increases, carbon dioxide assimilation and plant dry weight increase. For this reason, increases in plant dry weight when plotted against intercepted solar radiation commonly show a linear relationship. For example, the linear relationship found through the growing season for four cultivars of peanut is shown in Fig. 5.2. The linear slope of this plot has been defined as the 'radiation use efficiency' or RUE. The units of RUE are grams of accumulated plant mass per MJ of intercepted solar radiation.

Not surprisingly, however, the value of RUE varies among species and with environmental conditions. The main factor influencing variation in RUE is the photosynthetic capacity of the individual leaves. Increasing leaf photosynthetic capacity results in an increasing RUE (Fig. 5.3). Therefore, C_4 species (e.g. maize) with higher photosynthetic capacities than C_3 species (e.g. rice and soybean) have higher values of RUE. For many non-stressed situations, the RUE of maize is about 1.6–1.7 g MJ^{-1} in contrast to rice and wheat that commonly have a RUE of about 1.1–1.4 g MJ^{-1}.

Shifts to low leaf photosynthetic rates can be especially deleterious on RUE because only small decreases in photosynthetic activity can result in substantial decreases in RUE (Fig. 5.3). For example, decreasing leaf nitrogen content is closely associated with both decreased leaf photosynthetic rates and decreased radiation use efficiency. Drought conditions also result in decreased leaf photosynthesis rates and RUE.

Another factor that influences RUE is the amount of chemical energy it takes to produce different plant materials. Carbohydrates, including starch and cellulose have relatively low energy contents per unit weight. Therefore, those crops that produce seeds high in starch (e.g. wheat and rice) have higher RUE. Proteins and oils require considerably more energy to produce than

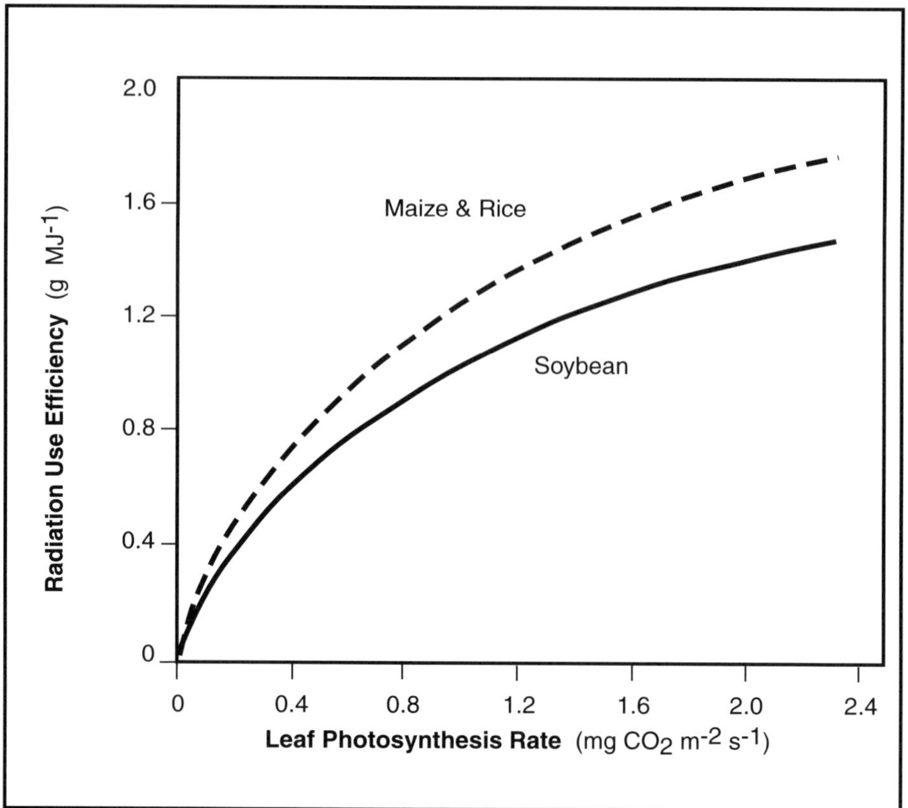

Figure 5.3 ***Theoretical relationship between radiation use efficiency and leaf photosynthesis rate for maize, rice, and soybean*** *(Sinclair and Horie, 1989).*

carbohydrates so that the amount of mass produced is less. Therefore, crops producing seeds high in protein and oil content (e.g. soybean and peanut) tend to have low RUE. The amount of seed mass produced per unit of photosynthate has been calculated for 24 species based on the biochemical composition of the seeds (Table 5.1). The seed production per unit of photosynthate for these 24 crop species ranged from a low of 0.42 g g^{-1} for sesame to 0.75 g g^{-1} for barley and rice.

An important fact shown by Fig. 5.3 is that at high leaf photosynthetic activity, further increases in activity lead to small increases in RUE. Even substantial increases in leaf photosynthetic rates lead to only modest increases in RUE, and consequently plant mass accumulation. The failure to sustain large increases in crop productivity by increasing the availability of fertilizer or by altering plant genetics, can be traced in part to the asymptotic response of RUE to increased leaf photosynthesis activity.

Table 5.1. Chemical composition, biomass productivity (grams of seed per gram of photosynthate), and nitrogen requirement (milligrams of nitrogen per gram of photosynthate) of seeds of 24 crop species (Sinclair and DeWit, 1975).

Species	Composition (percentage of dry weight)				Biomass productivity ($g\ g^{-1}$)	Nitrogen requirement ($mg\ g^{-1}$)
	Carbohydrate	Protein	Lipid	Ash		
Barley (*Hordeum vulgare*)	80	9	1	4	0.75	11
Bean, lima (*Phaseolus lunatus* mac.)	70	24	2	4	0.67	25
Bean, mung (*Phaseolus aureus*)	69	26	1	4	0.66	26
Chick pea (*Cicer arietinum*)	68	23	5	4	0.64	23
Corn (*Zea mays*)	84	10	5	1	0.71	11
Cotton (*Gossypium hirsutum*)	47	25	25	3	0.52	20
Cowpea (*Vigna sinensis*)	69	26	2	3	0.66	26
Flax (*Linum usilatissium*)	32	26	38	4	0.46	18
Hemp (*Cannabis sativa*)	27	29	41	3	0.44	20
Lentil (*Lens culinaris*)	67	28	1	4	0.65	28
Oat (*Avena sativa*)	77	13	5	5	0.70	14
Pea (*Pisum sativum*)	68	27	2	3	0.65	27
Peanut (*Arachis hypogaea*)	25	27	45	3	0.43	18
Pigeon pea (*Cajanus cajan*)	69	25	2	4	0.66	26
Popcorn (*Zea mays* var. *praecox*)	80	13	5	2	0.69	14
Rape (*Brassica napus*)	25	23	48	4	0.43	15
Rice (*Oryza sativa*)	88	8	2	2	0.75	10
Rye (*Secale cereale*)	82	14	2	2	0.72	15
Safflower (*Carthamus tinctorius*)	50	14	33	3	0.52	11
Sesame (*Sesamum indicum*)	19	20	54	7	0.42	13
Sorghum (*Sorghum vulgare*)	38	18	20	4	0.50	29
Soybean (*Glycine max*)	38	38	20	4	0.50	29
Sunflower (*Helianthus annaus*)	48	20	29	3	0.51	15
Wheat (*Triticum esculentum*)	82	14	2	2	0.71	16

Summation of Intercepted Solar Energy

The previous discussion described plant mass accumulation on individual days during the growing season of the plants. To understand the total mass accumulated by plants (B) during the growing season, it is necessary to sum the daily values for the entire season.

$$B = \sum_{n=0}^{d} I \times \text{RUE} \tag{5.3}$$

where d is the duration of the growing season. Equation 5.3 is deceptively simple. While RUE is generally stable through much of the growing season as discussed above, all three parameters on the right-hand side of the equation can vary. The environment, particularly temperature, has a great influence on both I and d.

During the later stages of seed growth, RUE has been consistently found to change in intensively managed crops including wheat, rice, soybean, and maize. The decrease in RUE is linked to decreases in leaf photosynthetic activity. In many crops, the bottom leaves lose their photosynthetic activity and then leaves successively higher on the plant lose activity. Loss of leaf photosynthetic activity during seed development is closely associated with the loss of nitrogen from leaves. The production of large seed mass requires the deposition in the seed of large quantities of nitrogen in most crops. Transfer of nitrogen from the vegetative components of the plant, including the leaves, results in the cascade of events that result in decreasing RUE during reproductive growth. This sequence of events in crop plants has been called 'self-destruction'.

The daily value of I is dependent on both the level of incident radiation (Io) and the fraction of radiation intercepted (F). Variations in Io are dependent on short-term changes in the weather (i.e. cloudiness) and location-dependent variables such as latitude, altitude, and climate. Variations in F are not unusual. These variations result from natural seasonal changes in ecosystem leaf area. This is certainly true in ecosystems dominated by annual plants where a new leaf canopy must develop each year.

Many important crops of agricultural ecosystems are grown annually and do not form a solid canopy until about 6 weeks or more after emergence. They also lose lower leaves as the crops approach maturity. The value of F is initially near zero when there are no leaves and increases to values over 0.9 as the LAI becomes large. Crops grown for hay (e.g. alfalfa) have even more changes in radiation interception because virtually the whole above-ground portion of the plant is harvested several times during the growing season. There is sufficient LAI left after harvesting to intercept only a small amount of light until there is regrowth from crown buds. Perennial crops and many forests essentially begin and end the season with a full canopy and, therefore, have a large potential seasonal yield advantage over annual plants.

The total duration of the season (d) is especially critical in determining the limit of total accumulated dry matter. Temperature is the most important factor defining d for many ecosystems. Temperature extremes define those portions of the year when plant production can occur. Virtually no plants can support photosynthesis when temperatures are below freezing. Consequently, in higher latitudes the duration of plant dry weight accumulation is set by the period between the last freeze in the spring and the first freeze in the fall. For many temperate crops, the duration is even more restricted because daytime temperatures need to be greater than about 10°C for the plants to sustain high rates of photosynthesis. Even in warm climates, cool temperatures may restrict plant dry matter accumulation. This is because the plants may be sensitive to temperatures as warm as 20°C. For example, in peanut, temperatures below 20°C are inhibitory to photosynthesis.

Not only may the total duration of the growing season be limiting, but the lengths of individual stages of plant growth are important. Temperature, again, has a dominating effect on the developmental progression of most plants (see Chapter 9). Commonly, as temperatures increase, the time required to progress

Table 5.2. Observed and simulated oven-dry maize grain yield, duration (d), mean daily temperature (t), and incident radiation (r) from simulated emergence to maturity (Muchow *et al.*, 1990).

Sowing date	Cultivar	Grain yield ($g\ m^{-2}$)		d	t	r
		Observed	Simulated	(days)	(°C)	($MJ\ m^{-2}\ day^{-1}$)
Katherine, Australia						
25 Nov. 83	Dekalb XL82	832	796	88	28.2	24.0
7 Feb. 84	Dekalb XL82	809	837	93	26.7	22.0
10 Oct. 84	Dekalb XL82	817	833	85	28.9	25.5
6 Feb. 85	Dekalb XL82	809	810	95	26.3	22.2
20 Aug. 85	Dekalb XL82	763	884	90	27.3	26.0
29 Jan. 86	Dekalb XL82	854	828	89	27.6	23.8
30 Aug. 86	Dekalb XL82	820	832	84	28.7	25.5
1 Feb. 88	Dekalb XL82	823	819	89	27.6	23.5
Gainesville, Florida						
26 Feb. 82	McCurdy 84AA	1038	976	115	23.3	19.2
23 Apr. 83	Pioneer 3192	867	840	98	26.0	20.9
Quincy, Florida						
24 Mar. 77	Pioneer 3369A	1073†	1121	112	23.6	25.8
		1225‡	1349			
30 Mar. 78	Pioneer 3368A	879†	849	112	23.6	21.0
		828‡§	1027			
Champaign, Illinois						
4 May 82	Agway 849X	1101	1115	126	21.5	19.9
12 May 83	Pioneer 3378	1080	1123	109	24.0	22.8
Grand Junction, Colorado						
22 Apr. 82	NK PX74	1734	1573	146	19.2	26.7
5 May 83	SX 5509	1516	1606	141	19.3	26.2
7 May 84	Funk G4507	1647	1672	138	19.8	28.3
5 May 85	Dekalb 656	1521	1799	143	18.9	28.2
28 Apr. 86	Dekalb 656	1483	1479	153	18.0	22.4

† six plants m^{-2}.
‡ nine plants m^{-2}.
§ 30% barren plants.

through each developmental stage in the plant life cycle is decreased. Therefore, higher dry weight accumulations and yields are associated with cool temperatures that lengthen the season. For example, in comparing maize yields among locations, the highest yields occurred in the cool temperatures of a mountain valley in Colorado and the lowest yields occurred in the hot temperatures of the tropics (Table 5.2).

For many crops the daylength also influences the length of the developmental stages (see Chapter 8). The primary influence of daylength is on the time to flowering. Of course, delayed flowering usually means an increase in the overall length of the crop growing season and an increase in plant mass accumulation.

Allocation of Plant Mass

Yield is generally determined by the fraction of the total accumulated plant mass that is in the harvested component, i.e., the economic yield. There are many plant components that may be harvested as economic yield, e.g., seeds, fruits, leaves (tobacco and tea), stems (trees, celery and sugarcane), roots, and tubers. The ratio of economic yield (Y) to the total accumulated plant mass (M) has been defined as the harvest index (HI).

$$\mathrm{HI} = Y/M \tag{5.4}$$

Therefore, combining Equations 5.3 and 5.4 gives the following expression for defining economic yield.

$$Y = \mathrm{HI} \times \sum_{n=0}^{d} I \times \mathrm{RUE} \tag{5.5}$$

Much of the significant improvement in modern crop yields has been associated with increases in HI. By developing plants of shorter stature and large grain heads the harvest index, and consequently yield, of many cereals has been substantially increased. The so-called 'Green Revolution' of the late 1960s and 1970s for which Norman Borlaug won the Nobel Peace Prize in 1970, was in large part associated with the development of widely adapted cultivars of wheat and rice that had significantly increased HI.

The importance of the increase in HI for wheat yields is illustrated in Fig. 5.4. Release of successive cultivars has not resulted in increased mass production. However, HI has increased dramatically from about 0.35 before 1940 to about 0.5 in 1980. The increase in HI was paralleled by an increase in grain yield.

Since about 1980, only minor increases in HI have been achieved. This may in part be because the eventual production of grains requires plants to invest a certain fraction of their accumulated mass in roots, stems, and leaves. In considering the basic structure of cereals, it has been estimated that the maximum HI that could be expected was approximately 0.6. Consequently, it appears unlikely that further major yield increases in cereals can result from further major increases in HI.

Many major crops other than cereals have also been bred for high HI values. For example, HI for soybean is now 0.45–0.55, for maize it approaches 0.5 and for potato HI is 0.8. Although these HI values may vary, within a genotype they tend to be stable over a wide range of environmental conditions. This stability in HI enhances the desirability of genetically establishing high HI in a cultivar.

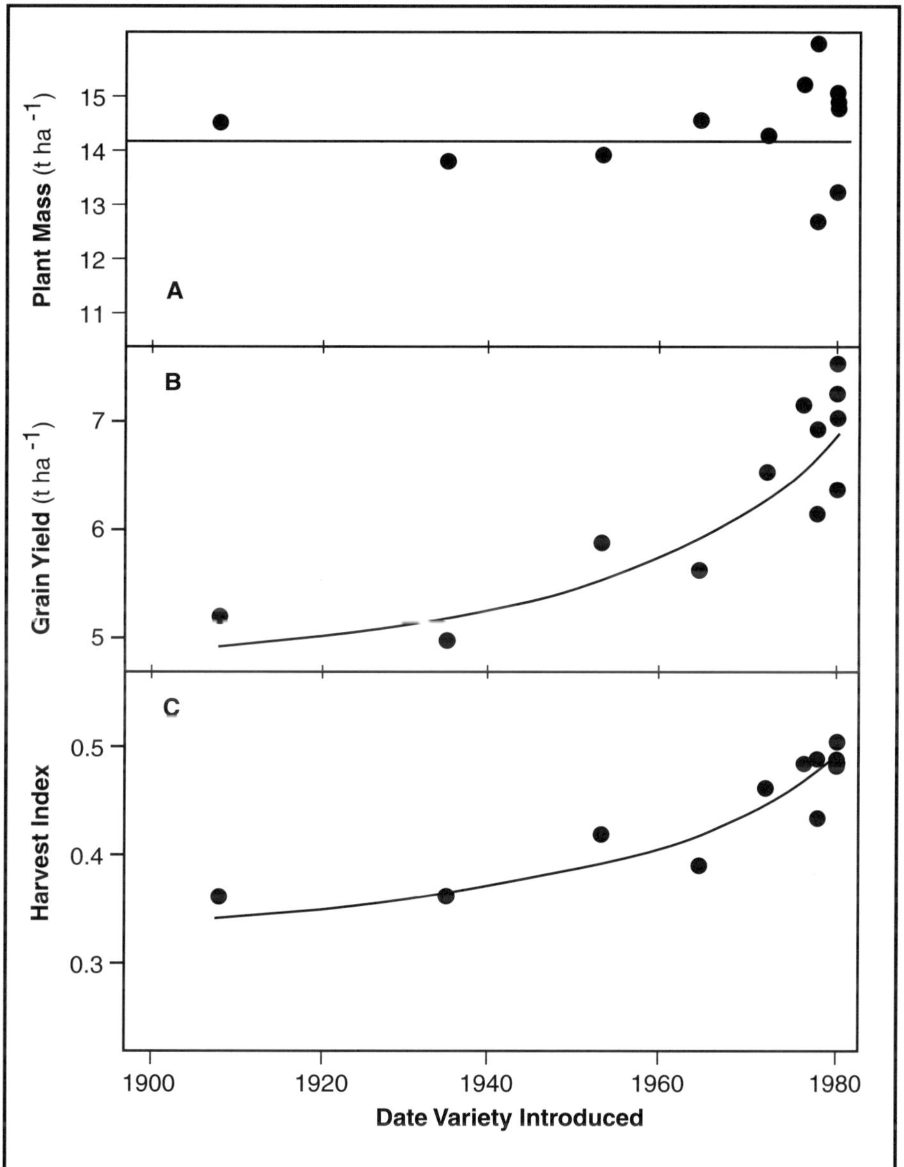

Figure 5.4 ***(A) Plant mass, (B) grain yield, and (C) harvest index of 12 varieties of wheat plotted against the year in which the variety was released to farmers*** *(Austin* et al., *1980).*

Stress Effects

Each of the four processes contributing to plant yield is sensitive to stress. Inhibition of the full expression of any of the processes results in a loss of potential yield. Already discussed has been the importance of temperature on leaf area development, photosynthesis, duration of individual stages of plant development, and total duration of growth. Extreme temperatures, particularly low temperatures, can have devastating effects on plant production. Large losses with major economic consequences have occurred in subtropical regions. The problem has been aggravated by the expansion of plant species that are sensitive to cool temperatures into regions that are infrequently subjected to cool temperatures. Examples include freeze damage on citrus in Florida and cool temperature damage on coffee in Brazil.

The consequences of inadequate nitrogen have also already been noted in several instances. Inadequate nitrogen can: (i) inhibit leaf development so that light interception is decreased; (ii) decrease leaf photosynthetic activity so that RUE is decreased; (iii) shorten the growing season as nitrogen is rapidly transferred from the vegetative tissue to the growing seeds; and (iv) decrease HI because required nitrogen is unavailable to continue seed growth. Usually, inadequate nitrogen will cause yield decreases by triggering all four of these negative events.

Inadequate water supply is the most common factor for decreasing plant productivity. Inadequate water can also decrease yield by decreasing the activity of each of the four main processes determining crop yield. Drought conditions directly influence leaf development by decreasing the final area to which individual leaves can expand. The decreased area of individual leaves means that the canopy leaf area index is decreased and light interception is decreased.

Inadequate water also results in a loss of leaf photosynthetic activity. During low soil water conditions, little or no carbon dioxide is assimilated by the leaves so that in these periods RUE is essentially equal to zero. Because water requirements are greatest at full canopy when photosynthetic activity is greatest, drought conditions are generally most common during the period of full canopy. Therefore, RUE may approach zero at exactly the time when maximum mass accumulation could be expected for the canopy.

Inadequate water may also alter the length of the growing season and HI. Severe drought conditions which threaten the survival of the plants can decrease the length of the season and the final value for HI. The imposition of drought conditions late in the season when water requirements are high, is most likely to result in premature crop senescence. This terminal drought late in the season can have especially devastating effects on HI and crop yield. Early crop senescence limits seed development and the production of harvestable yield.

Under conditions where there is inadequate water early in the season but water is available for normal crop maturity, the effect of the stress on HI is surprisingly small. Water availability late in seed development seemingly allows the normal pattern of seed growth to occur and HI attains values roughly equivalent to well-watered conditions. Of course, the effect of early season drought on yields may be substantial through the restrictions on leaf development and RUE during the drought period.

Other stresses can also be traced to their inhibition of one of the four major processes in yield development. Leaf area development can be restricted by direct damage from insects, diseases, and hail. Light interception can be restricted by direct competition with other plants (i.e., weeds). Leaf photosynthetic activity and RUE can be decreased by such factors as extreme temperatures and diseases. The length of the growing season can also be shortened by severe damage from many biotic and abiotic stresses.

Crop Yield Simulation

The four major processes describing the formation of crop yield provide a simple, but powerful approach for calculating or simulating crop yields. By assuming no stress conditions, the limit to plant production in various environments can be calculated. This approach was used in the assessment of maize yield potential among geographically diverse locations discussed previously (Table 5.2). The importance of temperature on the length of the growing season and the yield potential among environments was resolved in such a simple framework. Also, it was found that the yield potential for maize in each of these environments was matched by experimental observations. These results indicated that yield potential in maize may already be achieved based on current assumptions for maximizing leaf area development, RUE, growing season duration, and HI.

Similar analyses of environmental yield potential have been done for soybean and wheat. These analyses showed that a considerable amount of potential yield variation among environments depends on temperature and solar radiation by influencing the four processes describing the formation of yield. These studies emphasized the importance of temperature on influencing the length of the various crop developmental stages. Also, the calculated potential yield matches the observed yields in each environment.

Crop yield predictions have also been made for soybean, wheat, and maize for stress environments where there is inadequate water or nitrogen. Relatively simple functions can be attached to the basic, four processes to account for the stress effects. These model analyses have also been found to simulate the yields observed in environments with drought and/or nitrogen stress. These models provide a basis for examining potential managerial or genetic improvements in crops. They also provide a tool for yield prediction and environmental assessment on crop performance.

Comparison of Plant Production in Natural and Agricultural Ecosystems

Several attempts have been made to compare natural and agricultural ecosystem productivity but such comparisons are nearly always confounded by differences in the environment between the two systems. Furthermore, agricultural ecosystems are seldom more than a single species and are fossil energy subsidized

Table 5.3. Some high annual production rates for natural and cultivated systems (Loomis *et al.*, 1971).

Location	Vegetation	Net primary production ($t\ ha^{-1}\ year^{-1}$)	Daily growth rate ($g\ m^{-2}\ day^{-1}$)
Natural systems			
Europe	Spruce-pine forests	7–14	1.9–3.8
Former USSR	Birch forests	12	3.3
Europe	Beech	13	3.6
Former USSR	Grasslands	4–14	1.1–3.8
Composite	Subtropical forest	25	5.7
Africa	Tropical rainforest	30	8.9
Cultivated systems			
UK, The Netherlands and New Zealand	Ryegrass (*Lolium perenne*)	17–25	4.7–6.8
Southern USA	Bermudagrass (*Cynodon dactylon*)	20–30	5.5–8.2
Queensland, Australia	Various grasses	24.32	6.3–8.8
Caribbean and Hawaii	Various grasses	3–8.5	8.2–23.2
Hawaii	Sugarcane	42	11.5
California	Sugarbeet	30	7.7
California	Alfalfa (*Medicago sativa*)	20	5.8

whereas natural ecosystems are an unsubsidized mixture of species. If climates for the two systems are the same, agricultural ecosystems often do not use the full growing season as discussed earlier. Loomis *et al.* (1971) compared growth rates of the two systems on a daily and on a seasonal basis (Table 5.3) using data collected from several sources. The highest annual productivity was from sugarcane in Hawaii (4200 g m^{-2} $year^{-1}$) compared with a tropical rainforest in Africa (3250 g m^{-2} $year^{-1}$). The daily growth rates for the sugarcane and tropical rainforest were 11.5 and 8.9 g m^{-2} day^{-1}, respectively, assuming that both ecosystems had complete, year-long canopy coverage. The greater rate for sugarcane is consistent with the fact that sugarcane is a C_4 species and the tropical forest is composed mainly of C_3 species. Also, Hawaii is characterized by very high solar radiation for much of the year, while tropical rainforests are cloudy most of the time. Therefore, incident solar radiation is an additional cause of the yield differences between these two ecosystems.

Mitchell (1984) compared agricultural and natural ecosystems that had genetic, climatic, and edaphic similarities, namely, paddy rice with marsh vegetation (Table 5.4). Both environments had C_3 species and received natural fertility subsidies from flood waters, so water and nutrient deficiencies in both cases should have been small. The climate for both was similar. Annual aboveground production and growth rate for natural and agricultural ecosystems were similar.

Table 5.4. Rice productivity compared with natural marsh vegetation (Mitchell, 1984). Rice data are for single crops, and the maximum crop growth rates are for the dominant species while the average and annual above-ground TDM yield are for natural marshes, which had a mixture of species. The photosynthetic pathway (C_3 or C_4) is given when known.

	Growth rate (g m^{-2} day^{-1})		Total dry matter (t ha^{-1} $year^{-1}$)
	Maximum	Average	
Annuals			
Rice (C_3)	55	7–17	8–12
Zizania aquatica	–	21.6	15.8
Perennials			
Cyperus papyrus (C_3)	41	12–19	40–60
Spartina alterniflora (C_4)	64	2.2–15.5	5–40

Summary

Crop productivity is generally measured as the amount of plant mass accumulated into the harvested plant component of the crop during the growing season. The accumulation of harvestable biomass has been summarized as resulting from four processes.

1. The interception of solar radiation by the leaves of the crop provides the energy for all plant production. The changing size of the leaf area displayed by the plants through the season greatly affects the amount of solar radiation intercepted.
2. The efficiency in which the intercepted solar energy is used in photosynthesis to assimilate carbon dioxide determines the amount of accumulated mass. This efficiency, described as radiation use efficiency, is primarily dependent on the photosynthetic pathway of the crop species (C_3 vs. C_4), and the energy content of the plant mass being produced.
3. The time used for mass accumulation determines to a great extent the total amount of mass accumulated by a crop. A long-season crop in a tropical environment clearly has a greater potential to accumulate total mass than a species growing in a high-latitude region with a short growing season. To a large extent, temperature determines the length of the crop growth period, either by influencing the length of crop growth stages, or by ending crop growth through temperature extremes.
4. The fraction of the accumulated mass allocated to the harvestable component of the plant has a large influence on crop productivity. Much of the yield increase in the 'Green Revolution' in the 1960s and 1970s was a result of increased allocation of mass to the grain, i.e., increased harvest index.

By describing each of the above four processes, quantitative descriptions of crop yield are possible. These crop models allow much of the variation in the productivity of differing crop species and differing geographical locations to be explained. Studies of these processes also allow comparisons of the productivity of agricultural ecosystems with natural ecosystems.

Further Reading

Austin, R.B., Bingham, J., Blackwell, R.D., Evans, L.T., Ford, M.A., Morgan, C.L. and Taylor, M. (1980) Genetic improvements in winter wheat yields since 1990 and associated physiological changes. *Journal of Agricultural Science (Cambridge)* 94, 675–689.

Bennett, J.M., Sinclair, T.R. Ma, L. and Boote, K.J. (1993) Single leaf carbon exchange and canopy radiation use efficiency of four peanut cultivars. *Peanut Science* 20, 1–5.

Boote, K.J., Bennett, J.M., Sinclair, T.R. and Paulsen G.M. (eds) (1994) *Physiology and Determination of Crop Yield*. American Society of Agronomy, Madison, Wisconsin.

Evans, L.T. (1993) *Crop Evolution, Adaptation and Yield*. Cambridge University Press, Cambridge.

Fageria, N.K. (1992) *Maximizing Crop Yields*. Marcel Dekker, New York.

Kiniry, J.R., Jones, C.A., O'Toole, J.C., Blanchet, R., Cabelquenne, M. and Spanel, D.A. (1989) Radiation use efficiency in biomass accumulation prior to grain-filling for five grain-crop species. *Field Crops Research* 20, 51–64.

Loomis, R.S., Williams, W.A. and Hall, E.A. (1971) Agricultural productivity. *Annual Review of Plant Physiology* 22, 431–468.

Mitchell, R. (1984) The ecological basis for comparative primary production. In: Lowrance, R. *et al.* (eds) *Agricultural Ecosystems*. John Wiley and Sons, New York.

Muchow, R.C., Sinclair, T.R. and Bennett, J.M. (1990) Temperature and solar radiation effects on potential maize yield across locations. *Agronomy Journal* 82, 338–343.

Sinclair, T.R. and Horie, T. (1989) Leaf nitrogen, photosynthesis, and crop radiation use efficiency: a review. *Crop Science* 29, 90–98.

Sinclair, T.R. and deWit, C.T. (1975) Photosynthate and nitrogen requirements for seed production by various crops. *Science* 159, 565–567.

Spedding, C.R.W. (1975) *The Biology of Agricultural Systems*. Academic Press, New York.

Squire, G.R. (1990) *The Physiology of Tropical Crop Production*. CAB International, Wallingford.

Soil and Plant Production

6

E.A. HANLON AND F.M. RHOADS

Discussions in previous chapters have alluded to the importance of soils and soil fertility in plant production ecosystems. Soil is critical in providing essential resources for plant growth, such as mineral nutrients and water. Much of agricultural development (Chapter 4) has been associated with increasing the supply of nutrients in the soil. Also, degradation of soil by erosion and salinity has been important in determining the sustained success of various plant production ecosystems. Issues associated with soils and plant production are of even greater concern today as the demands for plant products are increasing.

Soil area available for crop production is declining rapidly in most countries of the world. As population increases, more land is converted to urban uses such as homes, businesses, and roads. It is not uncommon today to see individual shopping centers that cover 5–10 ha; some buildings cover one or more hectares each. Each kilometer of interstate highway may cover 10 ha or more. In addition, erosion is reducing soil area suitable for crop production. Eroded cropland is being converted to forest lands and pasture. Even if cropland area were to remain at its present level, the area of productive soil per person would continue to decline as population continues to increase. Because of limited soil resources, those involved in food production must use soil management practices and cropping systems that conserve and sustain soil productivity.

What is Soil?

Most of the Earth's continents are covered with soil. Because soil is ubiquitous, it is often ignored as a dynamic resource that can be changed by human actions. Soil should be viewed as a foundation of plant ecosystems because it is the most persistent of components within the system.

CAB INTERNATIONAL 1998. *Principles of Ecology in Plant Production*
(eds T.R. Sinclair and F.P. Gardner)

Definitions

The definitions of soil are as diverse as soil is itself. Definitions from knowledgeable sources usually contain the following central elements:

1. Soil is loosely consolidated material formed at the Earth's surface.
2. Soil forming processes include climate, organisms, topography, parent material, and time.
3. Soil can be composed of both mineral and organic materials.
4. Soil acts as a host to organisms.

From these points, soil is clearly not a static entity, nor is it homogeneous. Soil should be viewed as a dynamic structure that is constantly undergoing changes, both physical and chemical. These changes in turn affect the local environment through fluxes in microbial populations, plant species, and erosion rates. However, there is considerable feedback within the system, resulting in further modification of the soil.

It is because of scale that many of these changes are not readily obvious to humans. For example, changes in climate happen over a much longer period than a single human lifetime. Changes in microbial populations are not readily apparent to humans because of the relative difficulty of human observation.

The Role of Soil in Plant Production Ecosystems

Deep well-drained soils on flat landscapes are more suitable for cultivated crops that leave low amounts of plant residue like cotton, peanut, tobacco, and potato. Soils on gentle slopes are suited for maize, sugarcane, and grain sorghum. These crops require fewer tillage operations and leave moderate amounts of residue for protecting the soil from erosion. Steeper sloping soils are better suited to small grains, sod crops, and forests. The root systems of these plants hold the soil particles against erosion, promoting soil aggregation with improved water infiltration, and maintaining productivity.

Crop rotation was practiced more extensively before the widespread use of chemical fertilizers (see Chapter 4). In a rotation system, nitrogen was supplied to subsequent crops by the inclusion of legumes (they obtain nitrogen from the air in symbiosis with bacteria). Some deep-rooted crops, such as Bahia grass or Bermuda grass, can bring phosphorus and potassium up from greater depths making them more available to shallow-rooted crops that follow. Other crops like buckwheat and sweet clover can obtain nutrition from rather insoluble minerals, rock phosphate, for example. When these plants are subsequently incorporated into the soil, the nutrients are more available to crops that follow.

Widespread use of low cost chemical fertilizer during the past 50 years has made it more profitable to grow cash crops continuously on the same land. The considerable cost of establishing legume and sod crops has discouraged this practice. Reduced income while non-cash crops are growing in the rotation also makes it more attractive to practice monocropping of cash crops on the most productive soils. Availability of pesticides to control insects, diseases, and weeds also replaces some advantages of crop rotation.

There are many advantages in using rotations in cropping systems.

1. Deep-rooted forage crops improve soil structure and productivity. Poorly drained soils (soils through which water moves slowly) may be improved by growing deep-rooting crops like kudzu, alfalfa, sweet clover, or Bahia grass in a rotation system. The improvement results from increased water movement through channels left by decaying roots. Improved drainage in these soils allows the growth of crops that require more air in the root zone. More rapid water movement causes the poorly drained soils to dry out quicker in the spring. The earlier drying allows earlier sowing dates for higher yielding crops such as maize and soybean.
2. More vegetative cover results in less erosion and water loss. Proper crop rotation systems allow soils to remain more productive.
3. Weeds, insects, and diseases are easier to control. Weeds that are difficult to control in one crop are sometimes easily controlled in another. For example, directed spray herbicides in maize control weeds that cannot be controlled in soybean. Insects and diseases that attack one crop may not survive during the growth periods of other crops in the rotation.
4. Labor is distributed throughout a greater portion of the year.
5. Income sources are more diversified. The diversification of income lowers the risk of losses due to weather and pests.

How is the Soil Described?

Soil is a resource. As such, description of the soil, including both positive and negative attributes, is a method of taking inventory. Soil inventories are important to both governments and land owners because properly completed inventories can be used in many ways. Some examples of these uses include land use decisions, tax-based management planning, crop production potential, erodability, and building construction suitability. In the USA, the Natural Resource Conservation Service (formerly the Soil Conservation Service) performs this function. The outcome of their efforts has been the documentation of soils by county.

Inventories imply some systematic method of description and classification, a taxonomic system for soils. In the USA, soil taxonomy has undergone an evolutionary process. Early work by Russian soil scientists was adopted and greatly modified by American workers. This taxonomy was introduced in the USA Department of Agriculture Yearbook in 1938. Within the USA, the 1938 system was greatly modified up through the 1960s.

Table 6.1. An example of soil taxonomy. The state of Florida has recognized the Myakka soil series as the state soil. Myakka is classified as: sandy, siliceous, hyperthermic, aeric Haplaquods.

Category	Element	Description
Order	Ods	Spodosols
Suborder	Aquods	Aquic moisture regime, meaning that the soil is flooded for a portion of the year, coupled with the soil order
Great Group	Haplaquods	Haploid meaning simple or minimum soil horizon formation, coupled with the suborder
Subgroup	Aeric	Aeric meaning aeration
Family	Sandy, siliceous, hyperthermic	The three components define the textural, mineralogical, and temperature characteristics of the soil
Series	Myakka	The name of the town near the first description of this soil.

Soil taxonomy

In 1975, a new soil taxonomy was published that allowed a better understanding of many of a soil's properties simply by knowing its classification. The concept uses a taxonomy system composed of a series of formative elements whose definitions suggest selected properties of that soil. These elements are often formed from Latin or Greek words describing that particular soil attribute, but may also represent an abbreviation of other words. As these formative elements are added, one moves further into the classification system. Table 6.1 illustrates the hierarchial structure of this classification system for a particular soil series.

The primary advantage of this system is that knowledge about soil properties is conveyed within the formative elements. Learning the meaning of the formative elements allows soils to be described by proper assemblage of the elements. This method is far superior to the classification of minerals, each named for their discoverer, a notable location, or a well-known person of the day. None of the properties of a mineral are described within such a naming convention.

The hierarchical structure of this soil taxonomic system begins by classifying a soil into one of 11 orders (Table 6.2). The formative element indicating the soil order is used as the base to which other elements are added. The second level of classification, the suborder, consists of the formative element representing the soil order, and a second element describing a major attribute or characteristic of the soil. Addition of a third formative element will produce the classification description at the great group level. Descriptive elements used in the naming of suborders and great groups address specific characteristics of a soil. Consequently, they provide information concerning its formation, chemical or physical characteristics, important diagnostic soil horizons, or climatic conditions.

Table 6.2. The eleven soil orders and a brief description.

Soil order	General description
Alfisols	A soil within which aquic moisture conditions exist for some time in most years, and this aquic regime (saturated with reducing chemical conditions) affects the redox state of iron within one or more horizons.
Andisols	Soils that contain an andic horizon, which contains certain percentages of aluminum and iron and volcanic glass.
Aridisols	Soils that are dominated by dry and/or salty conditions, the desert soils.
Entisols	Soils that show little evidence of soil-horizon development.
Histosols	Organic soils that do not have andic horizon.
Inceptisols	Soils that are forming under leaching conditions, but retain some weatherable minerals.
Mollisols	Soils containing high proportions of calcium, magnesium, and other bases, with a mollic soil surface horizon, which is a layer high in organic matter with ≥50% calcium, magnesium, and other bases.
Oxisols	Soils formed within tropical or subtropical regions that are highly weathered.
Spodosols	Soils that contain a spodic horizon, which is an alluvial layer containing organic matter and aluminum with or without iron.
Ultisols	Soils with horizon containing appreciable amounts of clay, but with low levels of calcium, magnesium, and other bases.
Vertisols	Clayey soils that exhibit wide, deep cracks during part of the year due to drying conditions and shrinking of clay.

There are three additional levels of classification within this taxonomy. The subgroup is constructed of the great group modified by an adjective describing an important aspect of the soil. When a soil family is described, three additional adjectives are added to the subgroup. These adjectives usually delimit information about the particle-size class, the mineralogy, and the temperature regimes of the soil being described. Often adjectives used at the subgroup and family levels are derived from formative elements.

The lowest level of classification is the soil series, which is usually a place name next to the location where the soil was originally recognized. This place name does not convey any additional meaning to a person who has no other source of information about the classification of the soil.

Soil horizons

Often the soil-forming processes result in layers within a soil profile. These layers can be seen in vertical cuts through the soil (Fig. 6.1). The so-called diagnostic horizons are indicative of the soil's genesis and help to place the soil within the classification system. A diagnostic layer occurring at the soil surface is called an epipedon, while layers beneath the surface are called horizons.

These epipedons and horizons contribute to proper management of crops if their chemical and physical characteristics are considered. For example, all

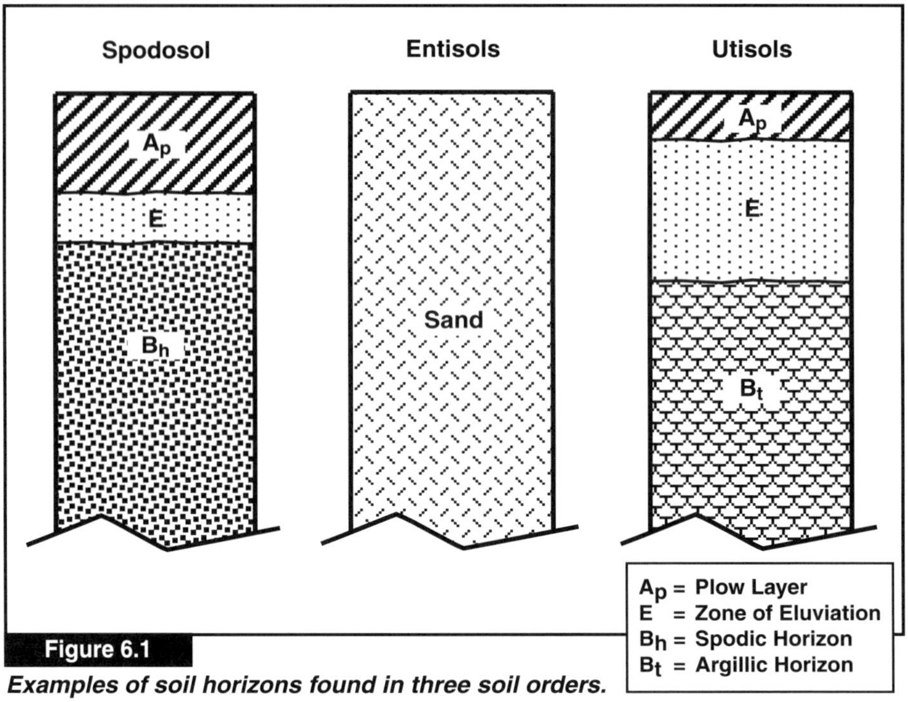

Figure 6.1

Examples of soil horizons found in three soil orders.

spodosols contain a spodic horizon (Table 6.2) that can be expected to be slowly permeable to water. Vegetables grown on spodosols in the central peninsula of Florida can typically be subsurface irrigated by induced or perched water table formed over the spodic horizon. Water is supplied to the crop by capillary rise from this artificial water table. This form of irrigation is inefficient when compared with microirrigation techniques because some of the subsurface irrigation water moves through the spodic horizon. Also, the spodic horizon will exist at different depths across a field causing spatially non-uniform availability of subsurface water.

Cation exchange capacity (CEC)

An important soil property that affects crop productivity and nutrient management is CEC. Certain minerals and organic matter often have an excess electrical charge due to flaws within the mineral crystal or unprotonated radicals. In most productive soils, there is a net negative charge resulting in the soil's ability to retain cations (positively charged ions) on these charged sites. The electrical charges are such that cations can be retained and released reversibly (Fig. 6.2). The ions of most concern are calcium, magnesium, potassium, and sodium, which are often called the base cations because of their basic reactions within the soil. Other ions of concern in acidic soils are aluminum, hydrogen (proton), and iron.

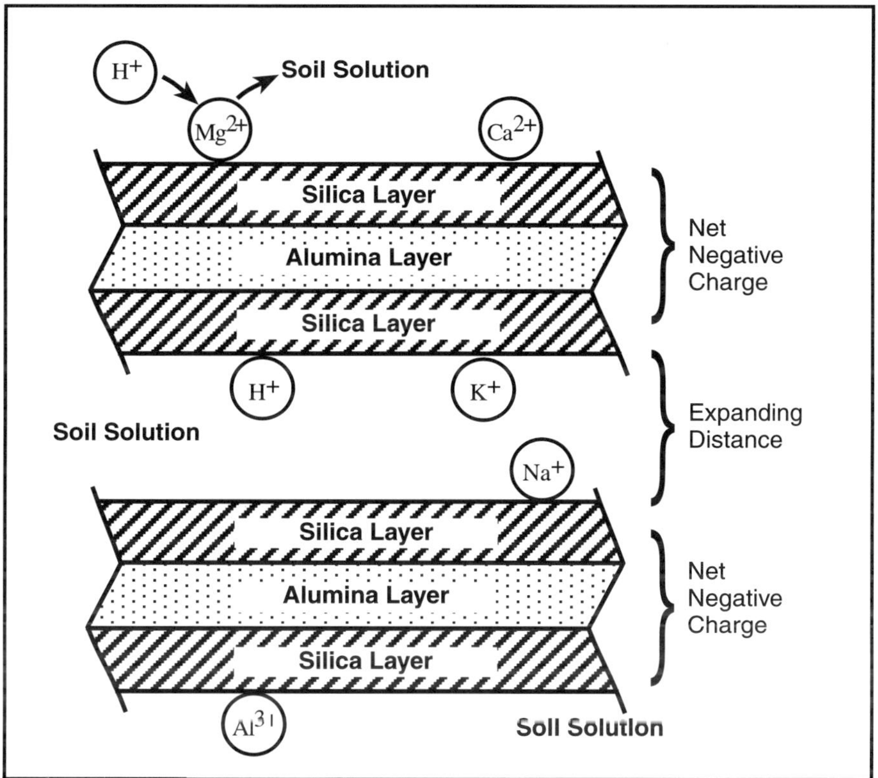

Figure 6.2 ***Schematic drawing of a cross-section of expanding clay soil particles with exchangeable cations.***

Ions are retained on these charged sites as a function of their concentration (chemical activity), and their ionic charge density (relative ionic radius as a function of charge). Ions with small ionic radius and high charge (e.g. aluminum or hydrogen) are held more tightly than ions with large ionic radius and low charge (e.g. potassium or sodium). Calcium and magnesium fall between these two groups with calcium being retained more than magnesium.

In sandy, low organic matter soils, potassium retention is small. This fact generally results in potassium being quite mobile compared with calcium, for example. Thus, potassium is often subject to leaching losses. As a rule, potassium is about ten times less mobile than nitrate, an anion not greatly retained by the soil's exchange complex.

Soil water

Movement of soil water is of interest when considering ground water contamination and nutrient leaching. Contamination and leaching are major problems for crop production on deep sands. Water moves rapidly through the profile of sandy soils because of their high infiltration rates. Therefore, sands in comparison to soils containing high fractions of clay, have small amounts of runoff and greater groundwater recharge. In terms of the movement of nutrients and pesticides, this means that sands have small amounts of runoff and large amounts of leaching.

To meet the transpiration demands of crop plants, soil must hold a significant amount of water available for plant use, against gravitational drainage, in its pore spaces. Large pores or macropores usually contain air except during or shortly after a rainfall or irrigation event, while small pores or micropores hold water against gravitational drainage. Plant roots require oxygen to function properly. Fortunately, soil typically contains about 50% pore space, only half of which is usually filled with water after gravitational drainage has occurred. The water content of soil when gravitational drainage is near zero is generally called field capacity. When plants wilt and fail to recover overnight, soil-water content is said to be at the wilting point. The wilting point is taken to be the water content when soil-water suction or matric potential reaches 1.5 MPa. The amount of water held between field capacity and wilting point is defined as the water available to plants.

For practical applications, field capacity of the root zone can be taken as the water content after 2 days of drainage following saturation. This definition is fairly accurate for coarse-textured soils, but finer-textured soils continue to drain for several days. The drainage rate of finer-textured soils after 2 days is, however, generally insignificant compared with transpiration rates of most crops.

Temperature

Besides soil texture and clay mineralogy, average soil temperature at a depth of 50 cm has been selected as one criterion in classifying soils at the family level. The temperature dependence of several soil properties has been shown. Soil colors tend to become more reddish and less gray with increasing temperature. Bases are more completely leached in warm areas, but rainfall is also important in the leaching of bases.

Soil texture

The fine-earth fraction of the soil (<2 mm) is used to describe a soil's texture. Somewhat arbitrary size divisions of the soil particles are made to separate the fine-earth fraction into sand (0.05–2 mm), silt (2–50 μm), and clay (<2 μm). When expressed as percentages, the textural class of the soil can be determined using Fig. 6.3. The size divisions presented herein are those used for soil taxonomic consideration, differing from the dimensions used for engineering purposes.

Soil textural classes are not affected by the organic matter content of a soil, since only the mineral fraction is considered to determine soil texture. Organic matter may influence the taxonomic classification of a soil, and does directly affect the ecology, but does not change the textural class.

Soil structure

Soil structure is the formation of larger aggregates from individual particles held together by a cementing agent. Cementing agents may be colloidal humus, clay particles, or metal oxides (i.e. iron and manganese). Individual aggregates are called peds and fit together along planes of weakness.

There are three features used to define soil structure. They are: (i) grade; (ii) class; and (iii) type. Grade refers to strength or resistance to shattering of peds and is moisture dependent. A weak grade of soil structure shows observable peds that are destroyed when removed from the soil profile. Moderate grade peds can be removed for careful examination in the hand, while strong peds are rigid and durable. Structural grade becomes stronger with loss of soil moisture, so the description must be defined in terms of moisture content such

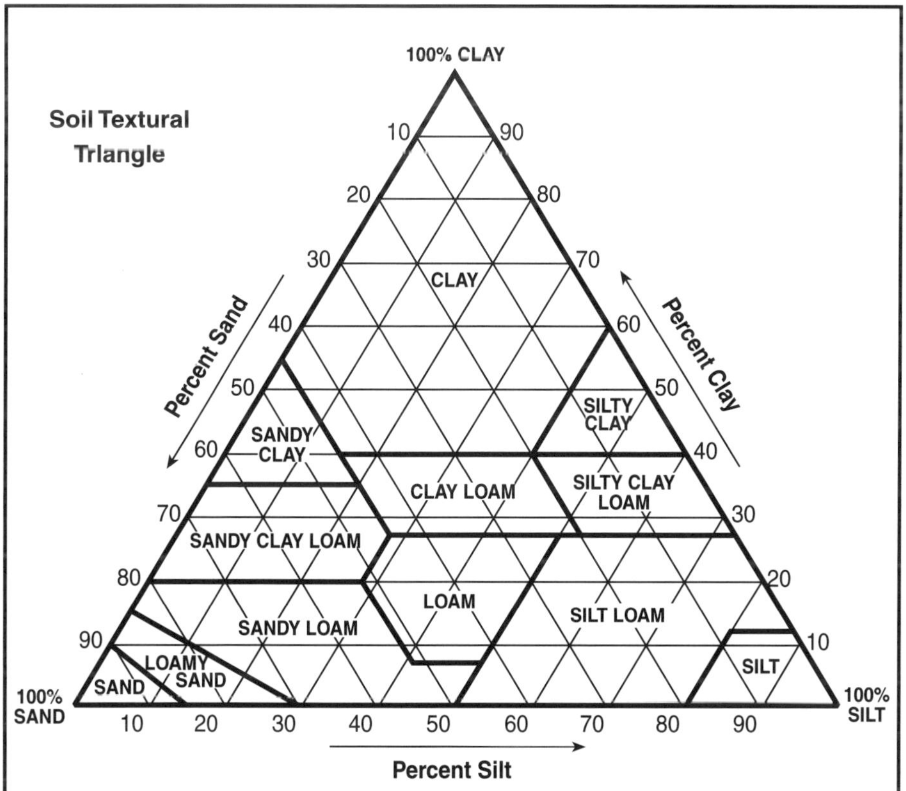

Figure 6.3 ***Soil texture classes based on the particle sizes.***

as weak when moist, strong when dry. Class of soil structure refers to size of peds (fine, medium, coarse, thick, thin, etc.). There are five types of soil structure: platy, prismatic, blocky, granular, and crumb; and two types of structureless condition, single grain and massive (individual particles adhere with no planes of weakness).

Soil structure is important because it influences pore size and distribution that control air and water movement (infiltration and permeability). When soil structure is destroyed by excessive tillage, water infiltration is restricted resulting in reduced water storage in the soil profile, increased plant stress, and lower crop yield. The inclusion of sod crops in crop rotation systems contributes to the development of a crumb structure in the upper root zone. The crumb structure allows greater infiltration of rain or irrigation. Also, deep-rooted crops may create root channels, in soil horizons that are massive and structureless. Such root channels improve water movement and storage capacity, and contribute to deeper rooting of following crops in the rotation.

Soil–Plant Relationships

Interactions of plants with soil contribute greatly to the success or failure of the plant. The effects of the plant on the soil also further modify both the physical and chemical characteristics of the soil. Given enough time, these effects can be a strong influence on soil formation. If these interactions are positive, a fertile and productive soil is usually the result.

Nutrients and plant roots

Most of the nutrition required by higher plants originates from the soil. The mechanisms by which nutrients in the soil are accumulated by a plant are complex, but a general understanding will aid in the consideration of plant production. Roots provide the nutrient collecting surfaces through which nutrients are accumulated by the plant. Not all roots are equal. Some roots possess additional structures that aid in nutrient uptake, such as root hairs. Other roots may have been sealed against uptake with plant-produced waxes and oils, usually as the root ages. Some general physiological conditions can be described that aid in nutrient uptake.

1. The more fibrous the root system, the more root surfaces are in contact with the soil solution and soil particles.
2. Those plants that have root hairs are usually more efficient in nutrient accumulation.
3. While a central large root(s) serves to stabilize the plant, its contribution to nutrient uptake is usually less than finer roots.

As roots grow, they expand in both diameter and length. Growth is also associated with full differentiation of cells within the root. These changes affect nutrient uptake. As the diameter of the root increases, its surface area also increases. This increasing surface area is presented to the soil solution and surrounding minerals resulting in more sorption area for nutrient uptake.

New soil volume is explored as roots develop and elongate. Therefore, roots may extend into the soil where the concentration of one or more nutrients is higher, permitting increased uptake. For some nutrients, such as phosphorus, the root must be physically close to the nutrient source for adequate supplies. Interception of nutrients by physical contact with soil particles may be important in some situations for adequate nutrient supplies.

For some nutrients, such as calcium and magnesium, new growth is needed to allow sorption into the root. These elements are taken with the transpiration stream into the root just behind the meristematic zone, but before the segment of the root where cells have fully differentiated.

Diffusion and mass flow

Once roots have been positioned for nutrient uptake, nutrients must be maintained at appropriate concentrations within the soil solution for proper plant growth. Two mechanisms are involved with the constant renewal of nutrients from the soil to the plant: diffusion and mass flow.

Nutrients depleted in the volume around the root through uptake are at a lower concentration than at some distance from the root. This concentration gradient induces nutrient diffusion towards the root. Potassium and phosphorus are often supplied to the plant via this process. However, the distance potassium can diffuse is often a factor of ten greater than that of phosphorus.

During transpiration, water is mechanically drawn into the root due to several forces. Some nutrients, such as calcium, magnesium, and nitrates, are carried to and into the root by this mass flow action. While there are many reports about the dominant control by diffusion or mass flow, it is likely that most nutrients are controlled to some degree by both processes.

Nutrient mobility in the soil

Each of the soil-supplied nutrients is controlled by its chemical reactions. Some nutrients are subject to chemical precipitation while others are retained reversibly on electrically charged sites in the mineral and organic soil fractions. Still others are not very reactive chemically, moving with the soil solution, subject to intervention from microbial activity or plant uptake.

These differences directly influence the distance from which nutrients may be supplied from the soil to the root, a result of the relative mobility of nutrients within the soil. Management of mobile nutrients should be different from immobile nutrients. Mobile nutrients are subject to leaching losses, but placement within the root zone for proper crop use may be less critical than for immobile

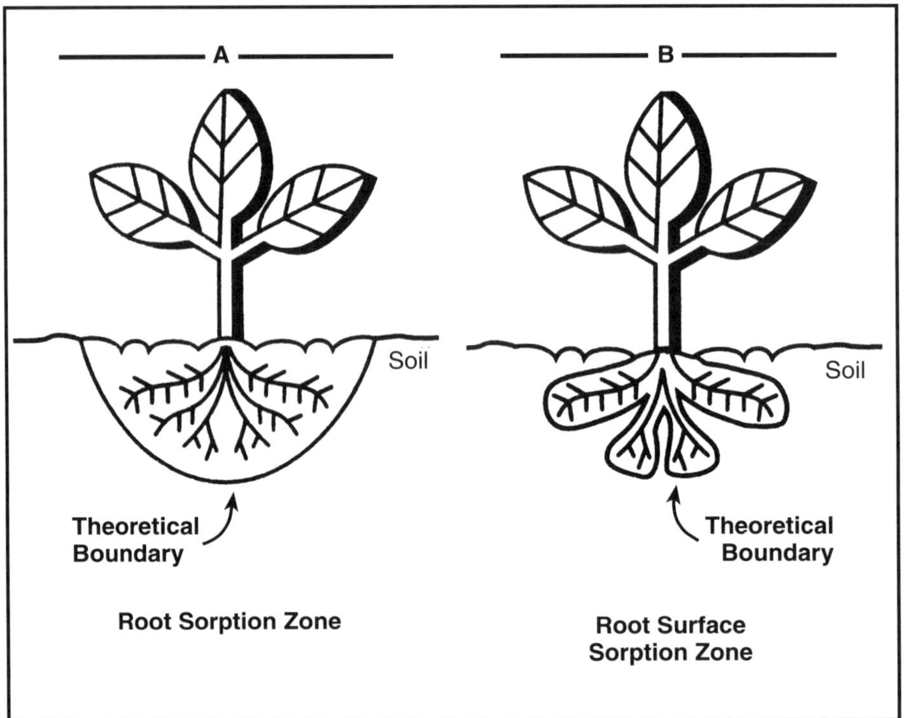

Figure 6.4 *(A) Root sorption zone and (B) root surface sorption zone in the soil.*

nutrients. Alternately, immobile nutrients are less subject to leaching losses, but may be moved off-site by soil particle erosion. Placement within the root zone may be more important for immobile nutrients to insure proper plant uptake. Placement must also address the reason for immobility. For example, if the soil is calcareous, calcium precipitation of phosphorus fertilizer is expected. Banding of phosphorus would delay this undesirable reaction, but still allow root sorption from the band. The band retards chemical precipitation by reducing the number of phosphorus granules that are in direct contact with the soil. Broadcast placement and incorporation of phosphorus in this case would address relative immobility allowing root access. However, the increased fertilizer/soil contact would greatly speed phosphorus reversion to plant-unavailable forms. Broadcasting would also require considerably more phosphorus fertilizer on a per area basis than banding to produce an equivalent plant response.

To aid in understanding the effects of mobility, two 'zones' or soil volumes can be defined. For the more mobile nutrients, such as nitrates, sulfates, and boric acid, a *root sorption zone* is defined. This zone is the volume of soil within which a mobile nutrient might be taken up by a plant's root system. This zone (Fig. 6.4A) extends beyond the actual root system of the plant because soil water is pulled to the roots bringing with it the more mobile nutrients.

The *root surface sorption zone* (Fig. 6.4B), describes a much more confined zone from which roots might extract immobile nutrients, such as phosphorus, potassium, calcium, or magnesium. For phosphorus, this zone may extend 2–3 mm from the root. For potassium, a more mobile nutrient (depending upon soil chemical conditions), the zone may extend 2–3 cm from the root.

These two zones should be considered as useful learning aids and not static volumes. Each nutrient will have different soil volumes controlled by the microsite soil chemistry, microbial and root activity, and the physical nature of the soil, as well. It is evident however, that when these zones overlap with adjacent plants, these plants are in competition for that nutrient. As the mobility of nutrients decreases, plant-spacing-induced competition decreases because of the small volumes associated with the root surface sorption zones.

Constraints Imposed by Soil on Roots

Roots, the primary gathering morphological structure for soil-supplied nutrients, water, and physical stability, are adaptable to a diverse range of soil conditions. Adverse soil conditions for plant growth originate from natural or human-induced causes. For example, a fine-textured soil such as a silty clay, requires much more care when deciding about cultivation conditions than a loamy sand. The *physical* attributes of the silty clay may be degraded by improper tillage methods or timing. Alternately, *chemical* attributes, such as pH, may limit plant production by adversely affecting root development.

Physical constraints

Soil compaction

Soil compaction occurs when tillage implements are pulled through soil having a moisture content higher than ideal for tillage operations. Continued tillage at the same soil depth season after season, even when moisture conditions are favorable, results in the formation of compacted soil layers. These layers at the base of the plow layer are several centimeters thick and are called plow pans, tillage pans, or traffic pans. Characteristics of a plow pan include close packing of individual soil particles, lack of defined structure, reduced pore space, high resistance to plant root penetration, hard when dry, and brittle when moist. Tillage pans can be shattered by deep tillage (subsoiling), but they often remain shattered only temporarily, because compact layers are unstable. Tillage reduction and growing of deep-rooted crops tend to improve soil structure.

Other types of pans are brittle pans, organic pans, and fragipans. Brittle pans are similar to traffic pans, but they are not caused by tillage operations. Organic pans are high in organic matter, dark in color. Fragipans occur below weakly developed B horizons and are very hard when dry. Brittle pans, organic pans, and fragipans develop during the process of soil formation and are called natural pans. Some deep-rooted crops can penetrate these natural pans but these pans are mostly not affected by tillage operations.

Chemical constraints

Soil pH

An excellent indicator of a soil's chemical status is soil pH. A soil's pH (a logarithmic measure of the concentration of hydrogen ions) is a result of many complex chemical reactions. Plant response, the types of invading weeds, and often the resulting productivity of the land can be directly linked to this indicator, pH.

In the southern USA, previously uncultivated soils tend to be acidic. When soil texture is dominated by sand-size particles, such as along the Coastal Plains, neutralization of excessive acidity is relatively inexpensive. On finer-textured soils, such as the Piedmont Ridge, much more acidity must be neutralized to increase soil pH.

Soil pH directly measures *active acidity*, that portion of hydrogen ions that is found in the soil solution, or is readily soluble. The *reserve acidity* is that portion of hydrogen ions that is not readily soluble or occupies cation exchange sites on clay and organic matter. In the two examples given in the previous paragraph, the sandy soil and the clayey soil may have the same soil pH. Their active acidity is the same. The reserve acidity of the Coastal Plains soil will be much less than the Piedmont soil, which contains considerably more clay than the sandy soil.

To adjust soil pH, both active and reserve acidity must be neutralized. Therefore, soil pH can be used to identify soils that may require pH adjustment for a selected crop. Soil pH, however, should not be the basis for deciding upon a liming rate, the lime requirement. Additional testing to include both active and reserve acidity is required.

At soil pH between 5.0 and 5.5 and lower, Al toxicity is often found for many agronomic crops. Aluminum solubility increases with decreasing pH, and increasing concentrations of Al ions in the soil solution adversely affects root growth. Some grasses, such as Bahia grass, continue to be productive below pH 5.0 and are tolerant of the elevated Al concentrations. The most important purpose of liming is to avoid Al (and Fe) toxicity. The decision to use lime should be based upon the selected crop, since tolerance to lime is also species dependent.

Nutrient availability and soil pH

Most soil-supplied nutrients are chemically affected by soil pH. The cations, such as Ca, Mg, Zn, Fe, and others, become more soluble as pH decreases. Anions, such as molybdates, become less soluble as pH decreases. Phosphorus solubility in soils is affected by pH in a complex manner: at low soil pH, P interacts with Al and Fe; at higher soil pH (>6.0), P interacts with Ca. Phosphorus is most soluble within a range of pH 6.0–6.5. Plant availability, however, depends on the plant species, not just chemical solubility.

Toxicities

Excessive levels of micronutrients in the soil may cause phytotoxicity in some crop cultivars. However, soil chemical properties control, to some extent, the threshold level of micronutrient required to cause toxicity in crop plants. As stated above, plant availability of boron (B), copper (Cu), iron (Fe), manganese (Mn), and zinc (Zn) is higher at low levels of soil pH. Therefore, phytotoxicity is most likely to occur in acid soils. Molybdenum (Mo) availability increases as soil pH increases, but Mo toxicity is rare. Toxicity from B and Cu is not likely to occur at soil pH above 6.5, while Mn toxicity is most likely to occur at soil pH below 5.5. Phytotoxicity from Fe and Mn is most common in poorly drained soils where oxygen content is low. However, excess moisture can cause Fe deficiency in calcareous soils. Zinc toxicity is rare, but it has occurred in peanut crops growing in former pecan orchards. It has also been observed following maize fertilized with high Zn levels at soil pH levels below 6.0. Deficiency of Zn can cause Fe toxicity in some plant species while Fe may interfere with Mn uptake. Thus, a second important aspect of controlling soil pH is to influence nutrient availability through control of nutrient solubility. Problems with both deficiencies and toxicities may be avoided by appropriate soil pH adjustment.

Selection of a soil target pH

Lime requirement is also dependent upon selection of a reasonable soil pH to which the soil will be adjusted, a *target pH*. Selection of a target pH for a specific crop has been a controversial subject influenced by both economic and crop management practices. If a crop is susceptible to Al toxicity, a reasonable lower limit of 5.0–5.5 will avoid this problem. Most species have a 1–2 pH-unit range that is considered adequate or optimal for plant growth. If this optimal range also includes pH 5.0–5.5, and no deficiencies are likely, then the target pH should be set within the 5.0–5.5 range.

Besides avoiding Al toxicity, prior crop management practices may dictate a different target pH. For example, on land in citrus production for considerable time, Cu toxicity has been reported. To avoid Cu toxicity and to continue to produce citrus, a target pH of 7.0 is recommended. Unfortunately, additional deficiencies, especially among the micronutrients, are often induced when soil is adjusted to this higher, but required, target pH.

Selection of a target pH should consider the crop, the desire to avoid both toxicities and deficiencies, cultural practices, and efficient use of liming materials. Often, the target pH will be set within the lower section of the crop's optimum pH range to address these potential chemical limitations controlled by soil pH.

Soil Testing and Fertilizer Management

As presented in the discussion on soil chemical problems, soil testing is a useful tool for management decisions. Use of soil testing is a regular activity when making decisions for fertilization in plant production systems. However, the underlying philosophy used to interpret soil-test results directly controls the forthcoming fertilizer recommendations. Factors to be considered in the soil-test interpretation and recommendation process include the efficiency of added fertilizer, its eventual fate and environmental consequences, and a positive crop response to fertilization.

Competing philosophies

Soil testing of croplands dramatically increased after World War II, when inorganic fertilizers became widely available. With low-cost fertilizers, crop responses were often well above traditional yields, and fertilization became an accepted practice. Recommendations were often based upon soil testing. In the USA in the 1960s and 1970s, commercial laboratories took a major market share of soil testing from the Land-Grant Universities. During those same decades, a marked shift in philosophies between commercial and University laboratories took place.

Basic cation saturation ratios

One of the first philosophies to be explored through research was the basic cation saturation ratio approach. An ideal ratio of cations (Ca, Mg, K, Na, and H) was thought to exist on the cation exchange sites within the soil. Adjustment of the ratios was accomplished by addition of fertilizer. Once the ideal ratios were attained, optimum plant growth and yield were expected.

This approach to fertilizer recommendations is still practiced today, dominantly in commercial laboratories. Often, extractable nutrient levels are assumed to be equivalent to exchangeable nutrients, and the ratios calculated from extractable levels. This approach has appeal in that nutrients would be 'in balance', avoiding 'imbalances'.

Research has shown that plants are excellent integrators of their surroundings, and that plants are quite productive throughout a wide range of ratios. In effect, no ideal ratio exists, and fertilization is often recommended where no positive plant response should be expected. Modification via fertilization of these ratios only directly affects the soil, not the plant.

Build-up and maintenance

In this approach, the soil is thought to act as a reservoir for plant-required nutrients. A soil-test critical value for each element is established, above which no response to added nutrients is expected. Fertilization is then used to adjust the soil-test value at or above the critical value, the build-up phase. While the period used for the build-up phase varies from one or two seasons to several years, the intent is to increase the soil-test measurable nutrient values.

After achieving a soil-test value above the critical value, seasonal additions of nutrients are recommended to sustain the soil test near the desired level, the maintenance phase. Therefore, each crop receives fertilization each season.

This method of fertilization has considerable appeal, since a manager would like to build up the fertility of the land, and to provide good stewardship. Sustaining the productivity of land is certainly a laudable objective.

The underlying assumptions are: (i) that the soil can retain higher levels of nutrients without losses to leaching or erosion; (ii) that the critical soil-test level has been based upon quantitative research; and (iii) that a soil receiving fertilization that season is somehow less productive than that same soil with a higher soil-test value. The Coastal Plains of the southern USA do not retain, nor allow, build-up of many nutrients due to their low cation exchange capacity and sandy textures. As with the basic cation saturation ratio approach, this idea promotes fertilization of the soil to achieve a nutrient change within the soil. The plant species is often not even considered. For these reasons, overfertilization and possibly adverse environmental impacts have resulted using the build-up and maintenance approach.

The build-up and maintenance and basic cation saturation ratio philosophies are occasionally combined. The combined approach is used to adjust the cations, increase soil-test values to critical levels, and maintain the soil-test results at these elevated levels.

Crop nutrient requirements

This philosophy has been accepted widely by the Land-Grant University Laboratories throughout the USA. The crop nutrient requirement concept is the same as the percentage sufficiency concept or the single limiting available nutrient concepts. The basic premise is that there is some total amount of a specific nutrient needed for a plant species to achieve optimum growth with high yield and quality. The source of that nutrition may be from the soil (first) or supplemented through fertilization using either inorganic or organic sources. If the crop nutrient requirements of a crop are satisfied by the soil, no response to added fertilization is expected, and no fertilizer is recommended. In cases where the soil supplies part of the crop nutrient requirement, fertilizer is used to satisfy the crop nutrient requirement, delivering the deficit portion. In effect, fertilization is used only where a positive crop response is expected.

Assessment of soil fertility is by a calibrated soil test. When a soil test is calibrated, its values can be interpreted and used to predict crop responses to added nutrition. Calibration data must be generated in the field for each specific crop (or crop type). The crop nutrient requirement concept potentially allows for greater fertilizer use efficiency. The greater efficiency is possible because fertilizer is added only to help the crop, not to change the soil-test values or other soil parameters.

A comparison with the other philosophies has proven that the crop nutrient requirement approach has several advantages. It is the means to improve fertilizer-use efficiency, to minimize the use of scarce natural resources, and to maximize the positive effects of fertilizer use without lowering yield or quality. Also, the environmental challenges associated with commercial agricultural activities are less daunting when nutrients are managed properly.

Soil testing uses

As environmental awareness evolves, trust in methodical testing of every aspect of the agricultural system is increasing, often to prove or justify management decisions to a wider audience. Reliance on soil testing has increased dramatically, especially for possible inclusion as a regulatory device. The reproducibility of soil-test methods, laboratory quality assurance goals, and instrumentation have advanced rapidly. Generation of field data needed to link soil testing, crop response, and amount of fertilizer application rates has not been actively pursued. The result is that soil test calibrations are often based on outdated research or represent undocumented experience of crop/soil specialists.

Soil-test-based fertilizer recommendations using the crop nutrient requirement concept has a proven history as a valuable decision tool. Soil testing should not be the only tool, or used as a regulatory agent, without substantial verification of the recommendations. Plant nutrient demand depends on many interacting agents, such as weather, soil fertility, water management, and tillage. Fertilization based on calibrated soil testing is an approximation of needed nutrients, but must not be viewed as an absolute. For example, rainfall events totaling more than 75 mm within 3 days are sufficient to remove a critical portion of mobile nutrients from a sandy soil. Depending upon crop development stage, supplemental fertilization should be considered.

Prefertilization sampling

The bulk of field research used to calibrate soil testing has been based upon soil sampling at the beginning of the growing season before spring fertilization. In this *predictive* mode, the soil test is used to prescribe fertilization for the upcoming growing season. Changes in amount of rainfall, temperature conditions, management, etc., will affect the growth of plants during the season, but after the predictive soil test. It is logical that changes to fertilization practices should be made during the season as adverse conditions are experienced. The use of fertilizer during the season should always be considered with respect to the stage of crop development. For example, it is unlikely that a positive crop response will be experienced in most annual crops if the crop has reached the reproductive phase. Therefore, leaching rains at the end of the season should not be a reason to apply fertilizer.

Diagnostic sampling

Soil samples collected when plants are actively growing are defined as *diagnostic* samples. Often, soil samples are collected to determine the cause of some adverse plant response (nutrient symptoms, lack of vigorous growth, or poor plant performance). Calibration of soil testing for diagnostic purposes has not been done. In all cases, laboratories apply the predictive calibration information to results, although the plant itself contains nutrient resources. For this reason, diagnostic testing should include cultural practices, tissue sampling from 'good' and 'bad' areas within the field, pesticide and other spray applications, crop development stage, and rainfall with irrigation records. Soil testing is only one source of information that should be considered in diagnostic work.

Tissue testing uses

Management of nutrition is a complex system of judgments. These judgments need to consider seasonal weather, soil conditions and possible leaching or denitrification, the crop's growing stage, and pest pressures. The plant is the integrator of these factors. Therefore, the plant itself is often tested. The challenge is to identify one or more plant parts that are highly correlated with yield, and reflect the yield response to added nutrition during that growing season. The most effective sampling time is often restricted to the vegetative stage of many crops. When the plant proceeds into the reproductive phase, nutrient demands of the plant are dramatically altered. Many nutrients are moved to the developing fruit at the expense of source tissues. Addition of nutrients to the soil during the reproductive phase is often not effective in altering yield or quality.

After appropriate plant parts have been selected (correlated with yield and quality), interpretation of the resulting nutrient concentrations for proper management decisions is required. Unfortunately, tissue testing is much like soil testing, providing more information to help with fertilization decisions, but not providing error-free decisions. To aid with this interpretation, several approaches have been documented. Each of these approaches carries with it certain assumptions that must be understood before acceptance.

The first approach to tissue concentration interpretations was the use of selected 'critical values'. This approach is similar to that used in soil testing. In theory, a certain tissue nutrient concentration is identified by field testing above which the crop yield and quality are considered optimum or maximum within the growing conditions. Concentrations below that level result in reduced yield. Because of the possibility of rapid growth during the vegetative phase, concentration of nutrients is often affected by the so-called dilution effect. Additionally, this approach must allow for any necessary changes to the critical value during the growing season to reflect maturation of the plant. For that reason, a single critical value for the entire life of the crop is usually not appropriate.

Multiple field tests for the same crop reveal a significant range of critical values. For that reason, the concept of 'sufficiency ranges' was introduced. Plant response to added nutrition when tissue concentration is within the *sufficiency range* is unlikely. Again, much like soil testing, a series of ranges was identified related to crop symptoms and tissue concentrations. A *deficiency range* identifies concentrations in which yield reduction is greater than or equal to 10%. Response to added fertilization by plants within this range is likely depending upon the stage of growth and seasonal conditions. The upper concentration within this range is often defined as the critical value. If plants test within the *luxury consumption range,* fertilization history should be reviewed and reduced in future seasons. Within this range, nutrient efficiency is not high and possible adverse environmental impact is increased. The *toxic range* identifies tissue concentrations that are so high that decreases in yield and quality occur.

To address the apparent deficiencies in the critical value and sufficiency range ideas, the Diagnostic Recommendation Interpretation System (DRIS) has received considerable research effort. Additionally, this approach requires considerable field-sampled tissue data to form the 'normative' ratios to which the current crop ratios are compared. Crop-nutrient concentrations are converted to ratios, such as N/P, etc. These ratios are then compared with normative ratios to identify ratios that are outside the normative range. Criticisms of this approach are that the idea requires a large source of measurements (often about 1000 tests or more), some ratios have no physiological interrelationship within the crop, and that there will always be one or more ratios that may be considered lower than all other ratios. In reality, the large normative group may give adherents of this method a better understanding of plant response. Some state laboratories have taken advantage of this fact and combined the DRIS system with a sufficiency range approach. Nutrients within ratios identified as outside the normative range are reviewed individually using sufficiency range concentrations.

Overfertilization and water quality

Tests of both soil and plant tissue are designed to aid in fertilization management decisions. The resulting recommendations are often controlled by other factors, such as philosophical approach, than demonstrated crop response to added nutrition. Therefore, fertilization recommendations are often inefficient, *i.e.*, amounts may be more than the crop nutrient requirements. The fate of nutrients added to the soil and not used by crops is complex. Factors that influence nutrient fate include the mobility of the specific nutrient, soil characteristics, volunteer weeds, and water movement.

For example, nitrate-N is quite mobile within the soil, moving just behind the soil solution wetting front. If leaching conditions exist, nitrates can be moved below the rooting zone. Chemically, nitrates are relatively inactive. Once below the depth where roots or microbes can take up the nutrient, nitrate will persist. It is for this reason that nitrate is often found within both surface and subsurface water sources. With the controversy over the regulatory selection of 10 $\mu g\ NO_3\text{-}N\ ml^{-1}$ as the tolerance level for human health, the concern is appropriate.

Soils managed to create a high level of fertility (nitrogen) through the use of large amounts of organic matter are also particularly vulnerable to nitrate losses. A widespread misconception is that organic matter is a slow-release nitrogen source, and therefore, does not produce excessive nutrients susceptible to leaching. To be of use to the crop, nitrogen must be in a certain chemical form, regardless of its origin from chemical or organic sources. If the soil is 'fertile', then nutrients are present for crop growth, and the potential to introduce nitrates into water bodies exists. Nutrients should be conserved, just as soil. Both are resources that must continue to benefit future generations.

Soil Erosion

Movement of soil particles from one location to another by water or wind is called soil erosion. Water erosion is dependent upon climate, topography, vegetation, and soils. The main element of climate that influences soil erosion by water is rainfall. Temperature, however, is important because it can change liquid water to ice.

Runoff depends on rainfall intensity and duration. Topography influences runoff by length and degree of slope. Vegetation reduces runoff velocity because it provides resistance to water flow. In addition, vegetation increases water infiltration into the soil because of root growth and other biological activities that increase soil porosity. Infiltration is the entry of water through the soil surface, while permeability describes water movement through the profile. Therefore, rate of both infiltration and permeability of the soil influence water runoff during a rainstorm. If permeability is greater than infiltration, where vegetation has stabilized soil porosity, then infiltration will remain relatively constant. If permeability is less than infiltration, however, the initial rate of water entering the soil surface will be reduced to the percolation rate governed by permeability.

Soil erosion by water not only depends upon runoff. The dispersive action and transporting power of water are also necessary for erosion to occur. Dispersive action of water is proportional to rainfall intensity, duration, and raindrop size, and influenced by soil texture and structure. Fine textured (clay) soils are more resistant to dispersive action of water than medium textured (silty) soils. Sandy soils have the largest particle size and require more rapid water movement for erosion to occur.

Soil dispersion reduces infiltration by plugging surface soil pores and by forming a thin layer with much reduced permeability on the surface. Initial infiltration rates may be high during a rainstorm but decrease quickly. Therefore, high intensity, short duration storms may not cause as much soil erosion as low intensity, longer duration storms.

Vegetation ameliorates soil dispersion by reducing the impact effects of raindrops and reducing runoff velocity. The presence of vegetation also counteracts the effect of long steep slopes. Such slopes tend to increase soil dispersion because of greater runoff velocity and length of time soil is exposed to dispersive action of water.

Erosion-control practices rely primarily on the following means of reducing erosion: (i) vegetation; (ii) plant residues; (iii) improved tillage methods; (iv) residual effects of crops in rotation such as grass and legume meadow; and (v) mechanical supporting practices.

Wind erosion

Wind erosion involves soil movement over great distances. Fine particles of soil were moved hundreds of kilometers during the dust storms of the 1930s in the

USA. Many hectares of cropland were depleted of their productive topsoil. Strong winds and bare soil are the main factors necessary for wind erosion to take place.

Three main factors of wind erosion are: (i) wind; (ii) nature of the surface; and (iii) soil physical properties. The wind speed must be great enough to initiate movement of soil particles. This threshold velocity for movement varies with particle size. Fine particles can become completely airborne while larger particles bounce along on the soil surface and dislodge other particles. Wind erosion is not limited to soil movement over great distances, but soil is often moved from one side of a field and deposited on another. Surface factors that influence wind erosion of soil are the amount of cover (vegetation or crop residues) and roughness. Minimal wind erosion will occur where the soil is well covered with vegetation. Unevenness of the soil surface caused by tillage reduces soil movement by wind. Ridge tillage is practiced in some areas to reduce wind erosion while crop plants are small and vegetative cover is not adequate. Aggregate size and mechanical stability are the soil factors mainly influencing wind erosion. Large soil aggregates that resist breaking into smaller aggregates reduce the erosive effects of wind.

Control of soil erosion

The loss of topsoil from cropland due to erosion reduces productivity in proportion to the amount of soil loss. Besides soil loss, plant nutrients contained in the soil are also removed. Not only do the losses of nutrients reduce crop production, it also contributes to reduced environmental quality. Soil sediment transported from fields to streams and lakes causes a reduction in water quality. As a result, fish production is decreased due to algal blooms resulting from nutrient build-up, and recreational activities such as swimming and other water sports are limited.

Crop rotation can reduce soil erosion by including crops such as legumes and grass meadows. These crops protect the soil from the erosive action of rain and wind for longer periods than clean-tilled crops. Crops that contribute to improved soil structure increase infiltration rates, which result in reduced runoff and reduced erosion.

Other practices that help control soil erosion are contour plowing, terracing, and strip cropping. Contour plowing consists of plowing in the direction perpendicular to the dominant slope. As a result, each furrow transports water off the field at a low rate with low erosive potential. This practice can be very effective in reducing erosion on gentle slopes but is not adequate for steeper slopes. Terraces are relatively large levees laid out on the slope contour. Large 'V' shaped basins on the upslope side of the levees are formed for transporting runoff from cropland to wooded or pastured area at a slow rate. Horizontal distance between terraces decreases as steepness of slope increases. If the vertical distance between terraces is too large, the volume of runoff water will be greater than the carrying capacity of the terrace. In this case, there is the possibly that large gullies may form.

Strip cropping consists of planting alternating strips of a sod crop (legume or grass meadow) with strips of clean cultivated crops (cotton, maize, soybean, etc.). Each strip is laid out on the contour. The sod crop catches the sediment from clean cultivated row crops controlling erosion over the entire field. Strips can be rotated to maintain as much uniformity of topsoil depth as possible.

Summary

Soil forms the basis for plant production. Since soil is the result of the five soil-forming factors, it is quite varied in its suitability for intensive cropping. Despite this heterogeneity, a useful classification system has been used to take inventory of this natural resource. While the classification system can be used for other purposes, it is valuable for understanding the reasons for management decisions that are site-specific.

The interactions between a crop and soil are governed by many processes. Understanding these processes aids selection of beneficial cultural practices, including tillage, fertilizer placement, and water management. Agricultural testing of soil and plant tissue is a useful tool for deciding upon and adjusting fertilization, but underlying philosophies should not be ignored. Efficient fertilization is the result of field research and intensive management of both fertilizers and water use.

Further Reading

Allaway, W.H. (1957) Cropping systems and soil. In: Stefferud, Alfred (ed.) *Soil, the Yearbook of Agriculture*. US Government Print. Office, Washington, DC, pp. 386–395.

Buol, S.W., Hole, F.D. and McCracken, R.J. (1980) *Soil Genesis and Classification*, 2nd edn. Iowa State University Press, Ames, Iowa.

Dahnke, W.C. and Olson, R.A. (1990) Soil test correlation, calibration, and recommendation. In: Westerman, R.L. (ed.) Soil Testing and Plant Analysis, 3rd edn. SSSA Book Ser. No. 3. Madison, Wisconsin, pp. 45–72.

Geraldson, C.M. (1977) Nutrient intensity and balance. In: Peck, T.R., Cope, J.T., Jr and Whitney, D.A. (eds) *Soil Testing: Correlating and Interpreting the Analytical Results*. ASA Special Publication No. 29, Madison, Wisconsin, pp. 75–84.

McLean, E.O. (1977) Contrasting concepts in soil test interpretation: Sufficiency levels of available nutrients versus basic cation saturation ratios. In: Peck, T.R., Cope, JT., Jr and Whitney, D.A. (eds) *Soil Testing: Correlating and Interpreting the Analytical Results*. ASA Special Publication No. 29, Madison, Wisconsin, pp. 39–54.

Moraghan, J.T. and Mascagni, H.J., Jr (1991) Environmental and soil factors affecting micronutrient deficiencies and toxicities. In: Mortvedt, J.J., Cox, F.R., Shuman, L.M. and Welch, R.M. (eds) *Micronutrients in Agriculture*, 2nd edn. Soil Science Society of America, Madison, Wisconsin, p. 371–425.

Olson, R.A., Anderson, F.N., Frank, K.D., Grabouski, P.H., Rehm, G.W. and Shapiro, C.A. (1987) In: Brown, J.R. (ed.) *Soil Testing: Sampling, Correlation, Calibration, and Interpretation*. SSSA Special Publication No. 21, Madison, Wisconsin, pp. 41–52.

Soil Survey Staff (1992) *Keys to Soil Taxonomy*. SMSS Technical Monograph No. 19, 5th edn. Pocahontas Press, Blacksburg, Virginia.

SSSA (1984) *Glossary of Soil Science Terms*. Soil Science Society of America, Madison, Wisconsin.

Tan, K.H. (1994) *Environmental Soil Science*. Marcel Dekker, New York.

Thorne, D.W. and Thorne, M.D. (1979) *Soil, Water, and Crop Production*. AVI Publishing, Westport, Connecticut.

Troeh, F.R. and Thompson, L.M. (1993) *Soils and Soil Fertility*. Oxford University Press, Oxford.

Waisel, Y., Eshel, A. and Kafkafi U. (eds) (1966) *Plant Roots: the Hidden Half*. Marcel Dekker, New York.

Water

7

T.R. SINCLAIR AND J.M. BENNETT

Properties of Water

Liquid water is an essential requirement for life. The existence of liquid water is a fortunate consequence of two features of our planet. First, the mass of Earth is sufficiently large that there is adequate gravity to retain gaseous water molecules (in contrast with Mars). Second, the distance from the sun is such that temperatures on Earth favor water in the liquid phase (in contrast with primordial Venus).

Physical structure of water

What is it about liquid water that makes it critical for life? It turns out that the simple structure of the water molecule is especially conducive for interaction with other molecules. The H–O–H molecule does not orient itself linearly, but rather the hydrogen atoms form an angle of 105° to each other (Fig. 7.1). This asymmetrical arrangement of the water molecule results in a polarity of charge across the molecule and exposes the oxygen atom. The oxygen 'side' of the molecule has a negative charge and the hydrogen 'side' of the molecule has a positive charge. Consequently, the exposed oxygen atom can readily form a hydrogen bond with a hydrogen atom in a neighboring water molecule.

The amount of hydrogen bonding among water molecules determines the physical state of water. For example, when nearly all the oxygen atoms in a collection of water molecules are involved in hydrogen bonds, water is a solid, *i.e.*, ice. As ice is warmed to 0°C, only about 15% of the hydrogen bonds are broken, yet water becomes a liquid. Because liquid water contains a substantial lattice of hydrogen bonds, the fluid is essentially incompressible. Incompressibility of water is an important feature of plant life. The structure of many plant tissues, especially leaves and stems of low growing plants, is maintained by the incompressibility of water. The depletion of water from plants is often visually identified as 'wilting'.

Especially important in life processes is the fact that water exists as a liquid over a wide range of high temperatures. Because high temperatures are required for many molecular interactions important for life, it is essential to have a liquid medium in which these interactions can occur. The hydrogen bonding between water molecules is highly energetic and only at the high temperature of 100°C does water boil.

Liquid water vaporizes at lower temperatures as individual water molecules overcome the attractions between neighboring molecules in the liquid phase and enter the gas phase. Still, considerable energy is required to separate individual water molecules from the liquid phase as they vaporize into the gaseous phase. At 25°C, 44 kJ per mole of water is required for the vaporization of water. Plant and animal systems have evolved to exploit the high heat of vaporization of water in various approaches to remove excess heat from organisms.

Water as a solvent

Asymmetry of the water molecule is crucial in its ability to act as a solvent for ions and polar molecules. Ions have electrical charges that cause the polarity of the surrounding water molecules to be oriented with respect to the ion. A negatively charged ion causes water molecules to surround the ion in layers with the hydrogen atoms of water oriented towards the ion (Fig. 7.2). A positively charged ion is surrounded by layers of water molecules with their oxygen atoms oriented toward the ion. An important consequence of surrounding each ion with layers of water molecules is that individual ions are separated in the water lattice and are 'dissolved'. Salt is readily dissolved in water as the oppositely charged ions are separated in liquid water, and each ion is surrounded by layers of water molecules.

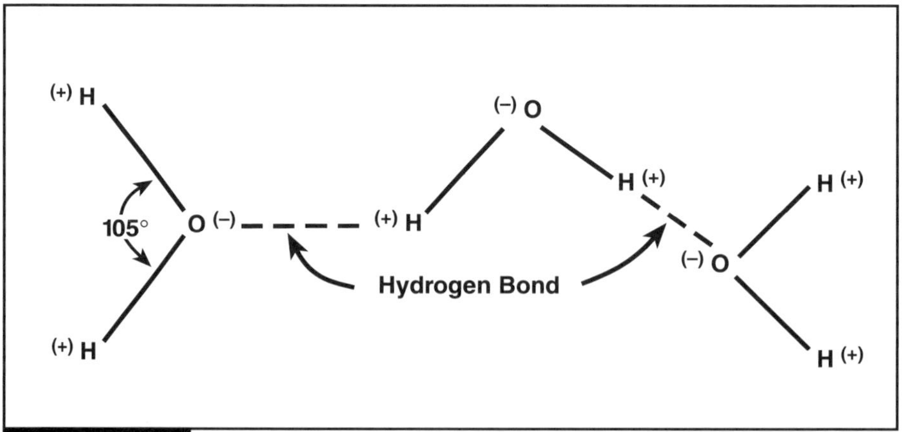

Figure 7.1 ***Schematic drawing of water molecules in the liquid state.***

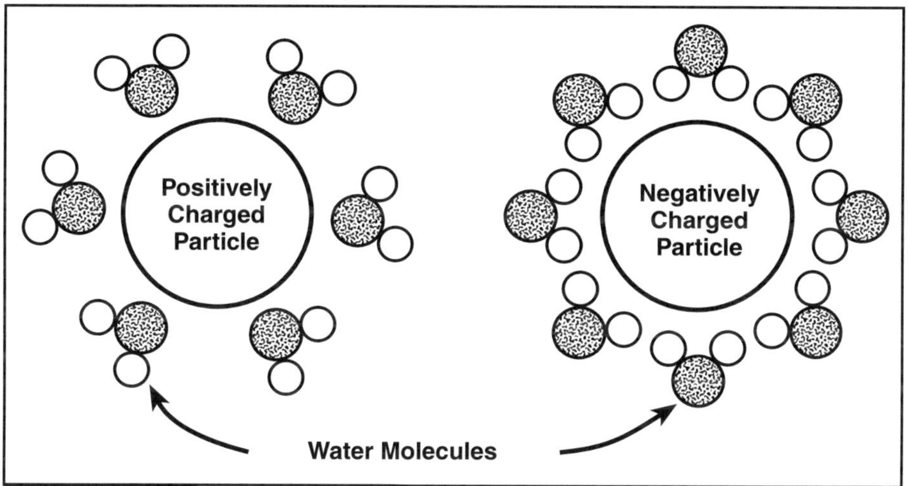

Figure 7.2 ***Schematic drawing of alignment of water molecules around charged particles.***

Small organic compounds such as sugars, which have localized sites of polarity, are readily dissolved in water. Larger molecules, such as proteins, become surrounded by shells of water molecules. These shells of water can be important in establishing protein structure and influencing the function of the dissolved protein. The dissolution of these organic molecules and the transmittance of polarity through the water lattice provides an essential medium for many biochemical processes of life.

The inability of water to interact with non-polar materials is also very important because it allows 'holes' to form in the aqueous medium for other biological functions. In particular, lipids are long molecules that are polar at one end and non-polar at the other end. The non-polar ends of adjacent lipid molecules are forced toward each other so that they are out of the lattice of water molecules. Membranes, which are an essential feature in nearly all living systems, are formed by this unique interaction of lipids with water. Two lipid layers orient their non-polar ends toward each other while their polar ends are dissolved in the water on either side of the membrane (Fig. 7.3).

Water Loss in Plant Production

Evaporation

Individual molecules within a liquid water lattice continually obtain enough energy to escape the liquid phase and enter the gaseous phase immediately above the water surface. The rate at which the individual water molecules escape the liquid phase increases with increasing temperature. Equilibrium is established when the number of molecules that escape the liquid phase is

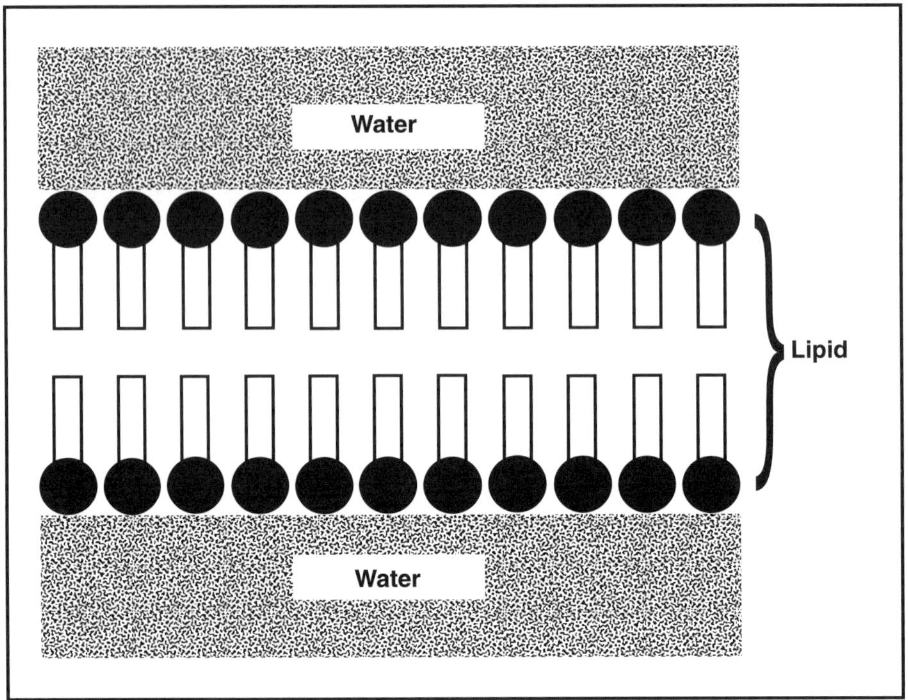

Figure 7.3 ***Schematic drawing of the alignment of a bilayer lipid membrane in water.***

balanced by the number returning from the gaseous to the liquid phase. The concentration of water in this air layer immediately above the liquid surface at equilibrium is called the saturated vapor concentration. Not surprisingly, the saturation concentration of water vapor increases rapidly with temperature because the escape rate of molecules from the liquid increases with increasing temperature.

The air layer immediately above the liquid water surface is saturated with water vapor. Consequently, the rate of water loss from a water surface to the atmosphere is dependent on the difference in water vapor concentration between the saturated air layer (p_v*, mol cm^{-3}) and the bulk atmosphere (p_v). Also, the conductance of the water vapor through the air above the saturated air layer (h, cm s^{-1}) directly influences water loss rate, or evaporation rate (E, g cm^{-2} s^{-1}).

$$E = 18\, h\, (p_v^* - p_v) \tag{7.1}$$

where 18 is the molecular weight of water (g mol^{-1}).

Equation 7.1 shows that there are three factors that influence the rate of water loss to the atmosphere. Usually the most important factor is the temperature of the evaporating surface because of the temperature effect on p_v*. The warm surface of a lake in summer results in a much higher p_v* and water loss rate, than the cool water temperatures in winter.

The value of p_v can also have major influences on E. The values of p_v are usually determined by general weather patterns so that p_v values remain within a fairly stable range for any given geographical region. In the dry air of a desert region where p_v is small, a considerably greater water loss rate occurs from liquid water than in a humid tropical region where p_v is fairly large.

Equation 7.1 also shows that the conductivity of the air is important in determining water loss. In the natural environment air is rarely calm so the value of h is determined to a large extent by the air movement above the evaporating surface. Increasing wind speeds lead directly to increasing values of h and consequently increasing values of E. Irrigation of a bare soil under calm wind conditions, usually existing at night, will result in substantially less evaporative loss than midday irrigation when wind speeds are usually greater.

Transpiration

Plant growth results from the assimilation of carbon dioxide (CO_2) from the atmosphere into organic compounds. The first step in this assimilatory process is the photosynthetic fixation of carbon dioxide inside leaves. For CO_2 to reach the photosynthetic apparatus inside leaves, CO_2 must diffuse from the

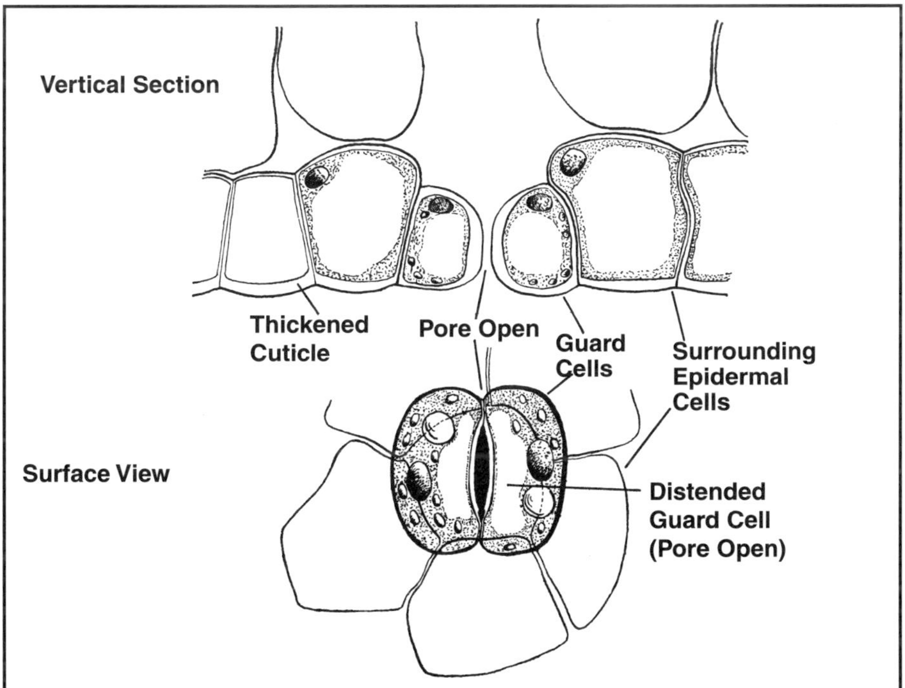

Figure 7.4 ***Drawing of a cross-section and face view of the guard cells of a stoma and the neighboring cells in a leaf.***

atmosphere to inside the leaves. The penalty of exposing leaf cells to the atmosphere for CO_2 assimilation is that the water-soaked cell walls are also exposed to the atmosphere. As a result, water is readily vaporized and diffuses to the atmosphere. The diffusion of gaseous water molecules from inside leaves to the bulk atmosphere is called transpiration.

A remarkable cellular structure has evolved to allow leaves to minimize water losses to the atmosphere because of transpiration. These structures, called stomata, are formed by at least two guard cells in the epiderma of leaves (Fig. 7.4). Between the guard cells is an air pore whose aperture is dependent on the turgidity of the guard cells. When conditions are favorable for photosynthesis, the stomata open to allow CO_2 diffusion into the leaf. Of course, the opening of the stomata also allows water vapor to diffuse out of the leaf to the bulk atmosphere. Under conditions not conducive to photosynthesis, such as at night, the stomata close and water loss is inhibited. Nevertheless, water is necessarily lost any time the stomata are opened to allow leaf photosynthesis. Loss of large amounts of water vapor from leaves to the atmosphere is an inevitable consequence of plant growth. These high rates of transpiration have profound implications for supplying adequate water to growing plants.

Defining the value of h for transpiration is substantially more complicated than for an open water surface because the conductance of water vapor molecules through the stomata must also be incorporated into h. The apertures of the air pores of stomata are highly variable and dependent on several physiological factors. There is no simple expression for quantitatively predicting stomata conductance. To complicate matters further, h must also include the conductance through the air surrounding the leaves. The complicated patterns of air movement around leaves define the value of this component of h.

Stomatal regulation of transpiration

Other than the environmental variables of p_v and temperature (leaf temperatures are generally near air temperature for defining p_v*), stomata are the key factors determining rates of water transport from leaves to the atmosphere. Those plants adapted to growing in environments with adequate soil water, commonly have stomata that have large apertures during the day to maximize CO_2 assimilation. This adaptation of open stomata has generally occurred in many agricultural crops because plant selections have been made to maximize growth and yield.

In view of the physics of leaf gas diffusion, the dual goal often expressed for plant selection of both high growth rates and low transpiration rates is impossible. To achieve high growth rates, stomata apertures must be open to allow rapid entry of CO_2 and its assimilation in photosynthesis. However, the fact that stomata apertures are open means that the conductance for transpiration is also high and the plant transpires large amounts of water. If water is to be conserved, then it is inevitable that the capacity to assimilate CO_2 must also be decreased by restricted stomatal conductance.

Essentially all plants adapted to humid environments where water is plentiful have large stomatal conductances to maximize growth. In addition, species in these well-watered environments commonly have values of stomatal conductance that are similar. Therefore, as a first approximation the transpiration rates of all closed plant canopies in a well-watered environment are equal. It is usually not the stomatal conductance that determines the total amount of water transpired under these conditions; it is the total energy supply to the canopy to support the vaporization of water. Therefore, when adjacent fields of soybean and maize both have sufficient leaf canopies to intercept most of the solar radiation, the transpirational water loss from the two fields will be approximately equal.

Similarly in natural, wetland ecosystems the replacement of one species by another species will usually have little effect on ecosystem water balance. The failure to understand the energy dependence of transpiration has led to confusion, for example, in understanding the problem of the invasion of Melalucca trees into the Everglades of Florida. These trees were introduced from Australia and it has been argued that they have resulted in a fivefold increase in evaporative water loss and a drying of the Everglades. Such a large change in transpiration rate is clearly not possible in view of the similarity in stomatal conductances among species, and the importance of the energy balance dictating transpiration rates.

Plants adapted to regions where episodes of soil water deficits frequently occur have evolved various leaf and stomata traits to cope with soil water deficits. These variations are common in arid environments where water conservation for plant survival is essential. Wax structures develop on leaf surfaces above the guard cells and decrease the conductance of gases external to the stomata. These wax structures attached to guard cells may be especially effective in restricting transpiration when the stomata are closed.

In arid environments the conductance of stomata can be further inhibited by placing the guard cells at the bottom of a fixed, small cylinder formed in the leaf epidermis. The addition of an extra length of still air external to the guard cells can greatly decrease the rate of water loss. This is true even when the aperture formed by the guard cells is fully open. Conductance is further decreased by the development of surface hairs that grow in the cavity above the guard cells and wax structures that cover the top of the cavity. Cacti and most other desert plants are prime examples of plants that have stomata adapted to water conservation at the expense of CO_2 assimilation.

Another approach to water conservation has evolved in a group of plants that have the capacity to assimilate and store CO_2 in organic acids at night (Crassulacean acid metabolism, CAM plants). The stored organic acids are then further assimilated in the photosynthetic metabolism during the following day. As a result, stomata are open only at night when p_v* is approximately equal to p_v so that transpiration rate is very small. Not surprisingly, these plants also have low growth rates because the capacity of CO_2 storage in organic acid is limited. Pineapple is the most important agricultural species that has this special photosynthetic sequence. Interestingly, in most commercial situations where pineapple growth is stimulated by irrigation, the plants commonly open their stomata during the day for greater CO_2 assimilation and transpiration rates.

Soil–Plant–Atmosphere Continuum

In the previous sections the factors influencing water vaporization and transpiration rates in leaves were discussed. However, an important question is, how does liquid water reach the leaves themselves? This question seems particularly challenging when the leaves may be as much as 100 m above the soil, for example, in giant redwood trees.

Water transport in plants

Hydrogen bonding among water molecules, as discussed previously, is a key factor in the movement of water through soil and in plants. The hydrogen bonding in liquid water results in a very strong attraction between water molecules and gives liquid water considerable cohesion. Some estimates place the cohesion of water at more than 30 MPa (1 atmosphere of pressure is approximately 0.1 MPa). Consequently, a water column can easily exist in a 100 m tree where the force of gravity is approximately 1 MPa.

Why does the water column flow from the base of the plant to the leaves? In developing the explanation of water flow to leaves, it is necessary to visualize the progression of water loss from leaves. At the end of the night it can be assumed that the cell walls in the leaves are bathed in liquid water. Once the stomata open and transpiration losses of water from the leaves begin, water is vaporized off the liquid surface on cell walls. Eventually sufficient water will be vaporized so that the surfaces of liquid water retreat into the very tiny pores in the cell walls. These pores are formed between the strands of cellulose and hemicellulose that make up the cell walls. These pores may have diameters of no more than a few nanometers. As the liquid water surfaces move into these pores, menisci form in the pores. Menisci are a direct result of the adhesive forces between the water molecules and the cell wall material and the cohesive forces among water molecules.

The menisci are indicative of a tension that develops in the water. This tension is transmitted through the column of water from the leaf cell walls, through the branches, stems, and roots, and finally to the water in the soil. In effect, the transpiration in the leaves creates a negative tension, or vacuum, that pulls water out of the soil through the plant up to the leaves.

Soil water

The response of soil water to the pull from plants is crucial in deciding plant behavior. Under conditions where there is considerable water in the soil, water flows freely to the roots and matches the demand for water resulting from transpiration. Under these conditions plants are rarely subjected to the stress of water deficits in their tissues.

As water is removed from the soil, however, important changes occur in the physical properties of the soil. Withdrawal of water in the soil occurs first from large pores and those pores become filled with air. Because water in the soil moves mainly in the liquid phase, the increasing cross-sectional area of air-filled pores means there is less area available for conducting liquid water. Consequently, the conductivity for liquid water in the soil decreases dramatically as the water content of the soil decreases. There may be a two to three orders of magnitude decrease in soil water conductivity as the soil dries in the range important to plants.

Water conductivity in soil is dependent on the amount of water in the soil. As a result, plant response to soil water is conveniently expressed on the basis of the soil volumetric water content. Volumetric water content is the ratio of the volume of water that is in a volume of soil (mm^3 water mm^{-3} soil). The upper limit of water available to plants is obtained by fully wetting the soil and allowing it to drain. This upper limit is commonly called field capacity. The absolute value of volumetric water content at field capacity varies a great deal among soils. The value ranges from 0.4 or more for clay soils, to less than 0.1 for sandy soils.

There is considerable ambiguity and controversy in defining the lower limit of soil water available to plants. Early attempts by soil scientists relied on a 'permanent wilting point'. However, experimental observations of permanent wilting point were subjective and the results depended on experimental conditions. Permanent wilting point is rarely used now because of these problems. More recently, the lower limit has been defined by the thermodynamic status of the soil water but this has also proved to be difficult to relate to plant responses. Plants do not respond directly to the thermodynamic status of the soil water. Using this approach has resulted in considerable ambiguity when relating the results to plant behavior. An emerging definition of the lower limit depends on quantifying the amount of water in the soil available for supporting transpiration. Consequently, the lower limit for available soil water is defined to occur when transpiration rates for plants on the desiccating soil are only small fractions of the transpiration rates of well-watered plants.

The total amount of water available to plants is the difference between the upper and lower limit. The interesting fact is that except for sandy soils, the amount of available soil water is roughly the same for most soils. That is, as a first approximation most soils provide about 0.13 available transpirable soil water on a volumetric content basis. Therefore, plants that can extract soil water from a depth of 1 m (1000 mm) have available a soil storage capacity of about 130 mm of water (1000 mm $\times$ 0.13 mm^3 water mm^{-3} soil). The expression of water storage as a water depth is a simplification in units from volume of water (mm^3) per unit land surface area (mm^2). Sandy soils have less available soil water, so they offer a smaller water storage capacity for plants.

The potential problem for plants in fulfilling their need for water can be examined in view of the available soil water. In a semihumid region at mid latitude the daily transpiration demand for water is roughly 6 mm per day. Dividing 130 mm by the daily transpiration rate indicates that the stored water in a soil that was initially fully wet would be exhausted in 22 days. In a semiarid region, the transpiration water demand is larger and might be 8 mm per day or greater. Assuming 8 mm per day, the stored water capacity would be exhausted in 16 days. These estimates show why substantial rains at least at 2–3 week intervals are required to maintain plant growth at a high level in most agricultural regions.

In the above calculations it was assumed that plant roots can extract soil water from a depth of 1 m. A 1 m depth of water extraction is reasonable for many soils and annual crops. However, it is easy to see that variation in the depth of water extraction is very important in determining the performance of a plant community. On soils that have fragipans or plow layers, roots may not be able to penetrate any deeper than 40 or 50 cm into the soil. It is not surprising that drought problems are frequently associated with these soils because of their shallower depth. On the other hand, deep soils and plants with deep-growing roots are much less prone to drought problems because of the greater soil water storage available to the plants. Perennial plants that can grow deep roots over several years may be especially able to use deeply stored water and are commonly observed to be more drought tolerant.

Plant Drought Stress

Plant responses

In the above discussion of available soil water it was assumed that all soil water was readily available to plants. This is not true. A decrease in soil water conductivity appears to limit the transfer of water to plants well before the lower limit of available soil water is reached. For nearly all plant species and soils, limitations to water transfer in the soil and stomata closure begin when approximately one-third of the available transpirable soil water still remains in the soil. Above the one-third level of available soil water most crop species show little, if any, inhibition of transpiration. When less than one-third of the available soil water remains in the soil, there is a steady decrease in transpiration rate with decreasing available soil water (Fig. 7.5). Therefore, water management schemes and irrigation schedules established to maximize yield need to provide water so that available soil water is always greater than the one-third level.

Once the one-third level of available soil water is reached, the water supply rate from the soil no longer matches the potential water loss rate through transpiration. Because the cells in the plant are essentially small, elastic water reser-

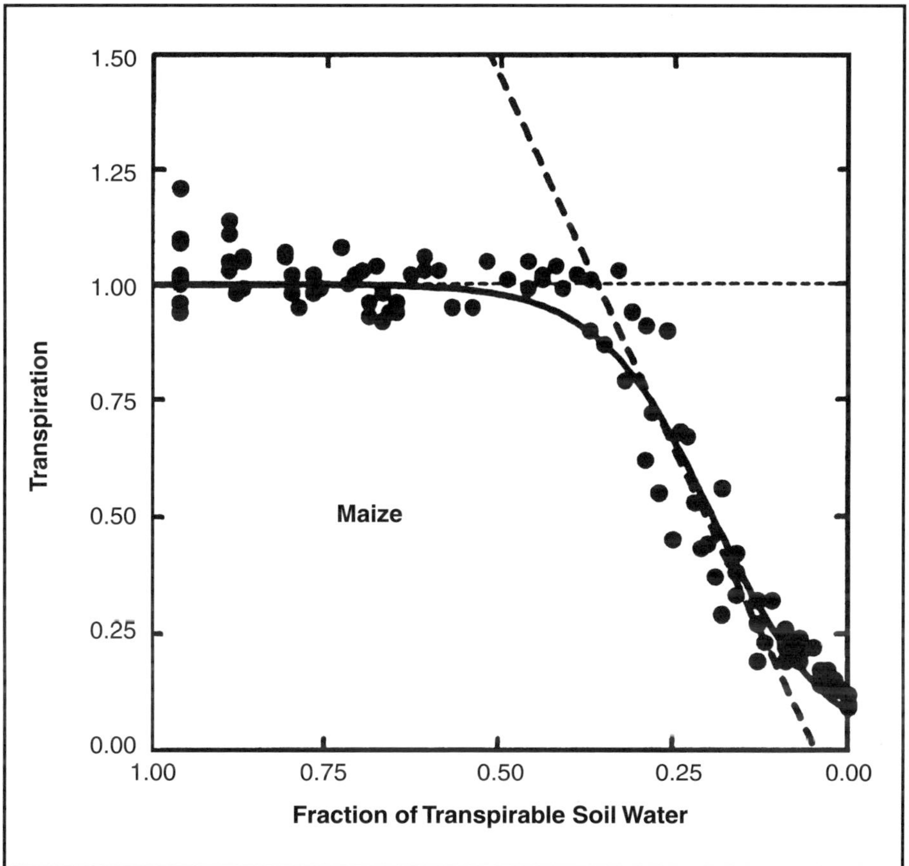

Figure 7.5 ***Transpiration rate of maize plants in response of soil drying from a wet condition (left-hand side) to a dry condition (right-hand side)*** *(Ray and Sinclair, 1997).*

voirs, when the water loss rate exceeds the supply rate there is a shrinkage in cell volume. This physical phenomenon is critical in the behavior of stomata guard cells because the loss of cell volume causes the size of the stomatal pores to decrease. The decrease in stomatal pore size results in a restricted transpirational rate that more closely matches the water supply rate. This physical adjustment in pore size to match water loss rate with the supply rate results in the steady decline in transpiration shown in Fig. 7.5.

Shrinkage of thin-walled cells because of an imbalance between water loss and supply during periods of low levels of available soil water, influences many processes in plants. Importantly, expansion and extension development of plant tissue is dependent on cell enlargement. If the water balance of the tissue is causing cells to shrink, or at least not to enlarge, then development of the tissue is arrested. Therefore, the decreases in the development of tissue associated with water deficits are also closely linked to available soil water. The decrease in the development of maize leaf area also begins to decline at about one-third available soil water similar to the transpiration response. However, leaf expansion declines more rapidly than transpiration in response to declining available soil water, apparently because of greater sensitivity in cell expansion associated with the declining water supply.

Especially important in some plants is the effect of inadequate water supply on the development of reproductive tissues. Again, the development of the reproductive tissue requires the enlargement of many cells, usually in a short time frame. An imbalance between water loss and supply at this time can severely inhibit reproductive development. A classic example of this problem is the development of silks in the female organ of maize. Because the silks must elongate by many millimeters to emerge from the ear, inadequate water supply can greatly delay silk emergence. Pollen production does not have a parallel elongation of tissue and pollen may be shed before the silks emerge from the ear under drought conditions. The level of embryo fertilization in maize may be drastically decreased in a high drought-stress environment.

However, yield levels in many crops cannot be improved simply by eliminating the sensitivity of reproductive development to drought. The basic limitation is the failure to accumulate carbon under stress conditions. The loss in reproductive development parallels the loss in photosynthate accumulation. As a result, the balance between grain production and the total biomass of the crop is frequently unchanged because of drought.

Drought tolerance

Drought tolerance is usually defined as the ability of cells and plants to maintain physiological activity at low water contents. This is an important physiological adaptation in many native plants in arid environments. These plants have developed mechanisms by which tissues and/or cells can tolerate desiccated conditions, and maintain or quickly recover physiological activity. These mechanisms are especially important for plant survival (Fig. 7.5).

In reality, tolerance mechanisms probably have little place in crop production except in arid environments of low production. If drought is sufficiently severe to regularly threaten crop survival, it is likely that the general poor water supply dictates low production. That is, regular exhaustion of the available soil water means that under these stress conditions no further increase in plant biomass and yielding capability can occur. Consequently, those regions that are regularly subjected to severe drought are inherently low yielding environments.

If drought survival is a recurring major problem for a particular crop in an arid environment, incorporation of drought tolerance in an existing crop is probably not useful. A better approach is likely to be alternate crops or plant production systems. Switching to crops or management systems with lower water requirements is likely to be the best option. For example, switching from maize to sorghum is a shift to a crop that has less demanding water needs (and with lower yielding potential). A switch to forage production using dryland grasses avoids the problem of drought survival in grain-producing crops.

Drought avoidance

Drought avoidance is a strategy by which desiccation of cells and tissue is avoided so that neither their activity nor survival is threatened. There are several approaches to drought avoidance that rely on management decisions and plant selection. In agriculture, a direct approach to drought avoidance is the selection of cropping systems that simply avoid the dry periods of the year. An important advantage of winter wheat production is the avoidance of the high water requirements of crop growth during the summer months. Also, short-season crops have been developed for drought avoidance. Short-season cowpea cultivars allow the life cycle of the crop to be completed within the time frame allowed by the available water. Consequently, the crop matures before or as the water supply is exhausted.

A direct approach for drought avoidance is the conservation of water by restricting transpiration. Some species have vascular systems in their roots or stems that allow only a maximum water supply rate to the leaves. When transpiration rate exceeds this supply rate, then a water deficit develops in the guard cells and the stomata aperture is decreased. Consequently, the water loss rate and water supply rate is brought in balance at a low transpiration rate. Because this decrease in stomatal aperture commonly occurs at midday when transpiration rates are potentially large, this phenomenon is called 'midday stomatal closure'. The importance of midday stomatal closure is that it conserves soil water for uptake by the plant at a later date. Of course, the midday closure of stomata results in a major restriction on CO_2 assimilation and plant growth rate. Many trees have midday stomatal closure at relatively low transpiration rates.

As suggested previously, selection of slow-growing plants is an important approach for avoiding drought stress. Compared with maize, sorghum has a slow production of leaf area so that early in the season before canopy closure it loses less water. Also, sorghum tends to have a lower leaf photosynthesis rate and stomatal conductance so that the water loss rate for sorghum may be less than maize. Therefore, sorghum conserves soil water and avoids droughts that may develop late in the life cycle of the crop. The conserved water may be especially important in an arid environment when a drought during grain development would prevent a non-conserving species from achieving seed maturity. Although no fundamental difference in response to drought exists between maize and sorghum, sorghum is universally acknowledged as a dryland crop because of its water conserving characteristics.

Additional drought avoidance mechanisms have been identified in other species. Pigeon pea, for example, abscises leaves as drought develops. By losing leaves the amount of leaf area for continued water loss is decreased and severe drought stress is avoided in the remaining plant tissues. Of course, the penalty of this approach is that upon rewatering, the leaf area for photosynthesis has been decreased.

Cowpea avoids drought stress by having stomata and leaves that are highly effective in sealing against further water loss once the available soil water is nearly exhausted. By inhibiting further water loss from the plant when the water supply is nearly exhausted, desiccation of tissue is avoided. This drought-avoidance trait has been effective in allowing cowpea to be grown in more arid regions.

Irrigation

An age-old solution to water deficits has been irrigation. Some of the earliest civilizations thrived because they developed ingenious irrigation systems. Irrigation continues to expand in modern agriculture. As discussed previously, the availability of water in the soil is essential to achieving high crop yields.

Irrigation water supplies

Many sources of water can be used for irrigation. The suitability of a water source is gauged to a large extent by the energy costs required to obtain and apply the water. Readily available surface water in lakes and ponds at the same elevation as the field means that pumping of the water on to the field is an inexpensive process. Of course, the costs of constructing dams and canals substantially increase the total irrigation costs.

Water that must be pumped to field height from wells, for example, is costly because of the energy expenses of elevating the water. Additionally, the recharge of water for the well may be an important consideration. Many irrigation efforts in arid environments depend on deep, ancient water reservoirs that are not being recharged or only recharged at a very slow rate. Consequently, the water resource for these irrigation schemes is finite and the depth from which the water must be pumped is increasing. This is the case of the Ogallala Aquifer in the high plains in central USA. The lowering of the water levels and the cost of pumping the water has discouraged increased irrigation in this region of the USA.

The oceans offer the greatest reservoir of water. Unfortunately, the salt content of sea water is much higher than can be tolerated by plants. (A few plant species grow in coastal regions because they have evolved special mechanisms to exclude salt during the uptake of water, or they may also excrete excess salt. These species are usually slow growing because of the necessity of developing special mechanisms to deal with salt.) To use sea water in crop production, it is necessary to remove the salt by reverse osmosis. This process requires considerable energy and is very expensive. Only in a few isolated regions of the world where oil is readily available as an energy source, is irrigation water for agriculture obtained from sea water.

Irrigation techniques

Having provided water to the field for irrigation, the next consideration is the efficiency in the application of the water. The lowest cost application procedure is generally by flood or furrow irrigation. Water is simply allowed to flow on to the surface of the soil and seep into the soil. Because additional pumping is usually not required in this application there are no further energy costs. However, the presence of a large surface of water allows high rates of evaporation directly from the standing water. The total amount of water lost during the irrigation depends on the amount of exposed water surface area and length of time of exposure.

High pressure distribution systems through nozzles or large volume 'guns' that spray the water over the field have been commonly used. While these systems allow for wide distribution of water and sometimes require less equipment than other approaches, there are some important disadvantages. A pump is required to pressurize the water distribution system. This extra pumping step can be quite demanding in terms of energy costs. As an alternative, systems have been developed that operate at lower pressures and decrease the energy costs.

Another disadvantage of the aerial distribution of irrigation water is that it can result in substantial losses of water during delivery. Irrigation systems that spray small water droplets can result in significant water losses. These losses occur because the droplets can be dispersed by the wind. Also, considerable evaporation from the droplets can occur in the air before the droplets reach the soil surface. Microjet irrigation schemes placed under the crop canopy near the soil surface help to decrease the loss of water in the air. However, many microjets are required per unit land area to distribute the water, so that costs may be prohibitive for all but the highest value crops.

Irrigation procedures that are receiving increased attention are drip and trickle systems. In these systems a tube is place on the soil surface or below the soil surface, and supplied with water at a low pressure. Either the water seeps through the walls of the tube directly (drip system) or special emitters are placed in the tube at regular intervals to release water (trickle system). Because very little of the soil surface is wetted with these systems, there is very little water loss directly to the atmosphere in these procedures. Low pressure water distribution also minimizes energy costs. However, the initial capital costs of these systems can be quite high. With little horizontal water distribution in the soil, the distribution tubes must be placed at narrow intervals. Once the tubes are installed in the field, their removal and reinstallation are expensive. Also, care must be given to supplying these systems with high quality water to avoid algae growth in and on the tubes. So far, drip and trickle irrigation systems are being used where water is very scarce or with high cash-value crops.

Irrigation issues

Application of water to agricultural lands raises several important issues. Irrigation in more arid regions where atmospheric vapor pressure is low means that evaporation is high (Equation 7.1). To maximize the efficiency in the use of irrigation water, priority for constructing new irrigation projects may need to go to regions with lower vapor pressure environments.

The high water loss rates coupled with the presence of salts in the irrigation water, results in the accumulation of salts in the soil. Over time, it is inevitable that salts will accumulate in the soil unless they are flushed out by over-irrigation or by rainfall. The build-up of salts eventually reaches levels that are deleterious to plant growth. One problem with some ancient irrigation schemes was soil salinization which eventually resulted in an agricultural collapse.

An idea to overcome the salinization problem is to develop crop plants that are tolerant to higher salt levels in the soil. Some success has already been achieved in identifying species and even some cultivars that can grow under somewhat saline conditions. The problem, of course, is that these selected genotypes are only a stopgap measure. As the land continues to be irrigated and salt concentrations increase further, these cultivars will eventually be affected by the increasing salt levels.

Ultimately, the solution to the salinity build-up in soils from irrigation is to flush the soil with water. Therefore, additional water needs to be applied to the irrigated land simply to carry the salts out of the root zone of the crops. Usually soil flushing requires substantial amounts of water that add significantly to the basic irrigation requirements.

In addition, salt flushing results in a major concern about the fate of the salt flushed from the root zone. Clearly, the water that flows away from the flushed land is very high in salt content and can be unsuitable for any other purpose. Therefore, special management procedures are required to dispose of the saline waste water from irrigated lands. Returning the water to a river that was the original water source degrades the river for uses downstream. This has been an issue in the Colorado River Basin because drainage water from various irrigation projects in Colorado, Arizona, and California is returned to the river. This drainage water results in a continual increase in the salinity of the Colorado River as it flows toward Mexico.

Simply building drainage canals to return the saline waste water directly to the ocean has also been found to cause environmental problems. For example, in northern California drainage canals have been closed because of the harm the saline water caused to natural estuarine habitats.

Although irrigation is an appealing solution to crop water deficiencies, the full costs of irrigation are rarely considered. The hidden costs include energy requirements, long term depletion of water resources, salt build-up in the soil, and the environmental consequences of saline drainage water. The regulation of water use in view of the water required by domestic and industrial water users is likely to alter the geographical regions where irrigation may be practiced.

Future emphasis on irrigation may move to the more humid regions where irrigation can be used as a supplement during critical periods of water shortage. Of course, irrigation in high rainfall areas will eliminate much of the concern for salt accumulation in the soil because rain water will flush and dilute the salt. Nevertheless, another set of important issues may develop for irrigation in more humid regions. Irrigation in a region subjected to intermittent rain storms can result in situations where a field is irrigated and then the irrigation is followed by heavy rainfall. The combination of irrigation and rain could result in significant problems of nutrient leaching, soil erosion, and local flooding because of saturated soils. Nevertheless, the potential benefit of limited irrigation during a critical time in crop production makes this an attractive management option.

Summary

The physical structure of water molecules confers many properties to water that make it essential for life. The polarity of the molecules means that they orient with respect to each other to form hydrogen bonds. These bonds allow water to exist as a liquid over a wide range of temperatures and at relatively high temperatures. Polarity of the water molecule allows it to dissolve the many charged ions and molecules crucial for life.

The loss of water molecules from the liquid phase into the gas phase occurs when individual molecules attain sufficient energy to escape the liquid surface. The movement of the water molecules from the air layer saturated with water vapor molecules immediately above the liquid, into the bulk atmosphere is evaporation. In leaves, water is vaporized at cell walls and moves through the stomata. This process is called transpiration. In general, under well-watered conditions, stomata are open and the transpiration rate for most species is roughly the same. However, as soils dry, the supply of water in the plant may cause stomatal closure and the rate of transpiration is restricted.

As soil dehydrates, the transport of water to the plant in the soil may become restricted. At this point water deficits in plants can develop resulting in loss of cell turgor and decreases in stomatal conductance. Many physiological functions in the plant may slow or be terminated depending on the level of dehydration in the plant. In agriculture, plant selection and management practices that avoid plant dehydration may offer the best opportunity for overcoming drought stress during plant production.

Irrigation has been used for centuries for overcoming drought stress in plant production. Some problems, however, are associated with irrigation. These include energy costs, capital equipment costs, adequacy of quantity and quality of water source, water losses in water distribution, and salinity.

Further Reading

Burman, R. and Pochop, L.O. (1994) *Evaporation, Evapotranspiration and Climatic Data*. Elsevier, Amsterdam.

Loomis, R.S. and Connor, D.J. (1992) *Crop Ecology: Productivity and Management in Agricultural Systems*. Cambridge University Press, Cambridge.

Muchow, R.C. and Sinclair, T.R. (1991) Water deficit effects on maize yields modeled under current and "greenhouse" climates. *Agronomy Journal* 83, 1052–1059.

Ray, J.D. and Sinclair, T.R. (1997) Stomatal closure of maize hybrids in response to drying soil. *Crop Science* 37, 803–807.

Sinclair, T.R. (1990) Theoretical considerations in the description of evaporation and transpiration. In: Stewart, B.A. and Nielsen, D.R. (eds) *Irrigation of Agricultural Crops*. American Society of Agronomy, Madison, Wisconsin, pp. 343–361.

Sinclair, T.R. and Ludlow, M.M. (1986) Influence of soil water supply on the plant water balance of four tropical grain legumes. *Australian Journal of Plant Physiology* 13, 329–341.

Stewart, B.A. and Nielsen, D.R. (eds) (1990) *Irrigation of Agricultural Crops*. American Society of Agronomy, Madison, Wisconsin.

Tanner, C.B. and Sinclair, T.R. (1983) Efficient water use in crop production: research or re-search? In: Waylor, H.M., Jordan, W.R. and Sinclair, T.R. (eds), *Limitations to Efficient Water Use in Crop Production*. American Society of Agronomy, Madison, Wisconsin, pp. 1–27.

8 Radiant Energy

T.R. SINCLAIR AND F.P. GARDNER

Sources of Energy

As discussed in Chapter 5, solar energy absorbed by green plants in natural and agricultural ecosystems is the essential, first step is sustaining nearly all life on Earth. Furthermore, solar radiation powers the weather systems. This includes temperature, precipitation and humidity, pressure systems and winds, and movement of both the atmosphere and ocean currents (such as the Gulf Stream). A long-term decrease by as little as 0.5% in energy output from the sun would cause drastic changes in climate, triggering another Ice Age.

In agriculture, crop growth is facilitated by large infusions of energy from fossil fuels. Of course, these fossil fuels are a result of solar energy absorbed by plants ages ago. Fertilizers, pesticides, mechanical power, artificial drying, and various other sectors of the agricultural infrastructure also require large inputs of fossil fuels. Consequently, the harvested plant material in an agricultural ecosystem can have less caloric energy than the fossil energy used in producing the crop. The input of fossil energy is so great that it has been said that modern societies are 'eating oil'.

Currently, there is interest in substituting photosynthate accumulated in plant dry matter for fossil fuels by producing alcohol or methane gas from plant material. Enthusiasm for the use of plant mass as energy sources increased sharply following the OPEC oil embargo in 1974. The technology is available, and the potential for high dry matter production in the long growing seasons of lower latitudes is large. This is especially true for C_4 grasses, such as sugarcane and Napier grass. Nearly all of the aboveground dry matter is harvested from these plants and put through a fermentation process to produce alcohol. In Louisiana, sugarcane yielded 1.2 MJ alcohol per MJ of fossil fuel energy required to grow and convert it to alcohol, even when sugarcane residues were burned to fuel the conversion process. Net energy yield was negative when fossil fuels were used to support the conversion process. The prospect for use of crop plants such as sugarbeet and fodderbeet in California is weaker than that

CAB INTERNATIONAL 1998. *Principles of Ecology in Plant Production* (eds T.R. Sinclair and F.P. Gardner)

of sugarcane. The energy gain in the production of alcohol from maize is positive, although about 8 ha of maize is necessary to fuel an average American's car for a year.

An important question in all schemes of using plant products for fuels is whether the small marginal gain in chemical fuels is worth the additional demands placed on ecosystems. To obtain these chemical fuels, ecosystems have to be managed to favor the desired species and to provide the resources to sustain high plant growth. As an alternative to large land areas devoted to producing chemical fuels in plant production ecosystems, natural ecosystems may be favored on a long-term basis. On the other hand, there is no real alternative to solar energy and plant production ecosystems in the production of food, feed, and fiber.

Solar Output

All materials with a surface temperature greater than absolute zero (0 K) emit radiant energy. Therefore, everything in the environment is emitting radiant energy. This radiant energy can be felt by your hand when it is held toward a warm object. The amount of energy emitted by a body is dependent on the fourth power of its temperature. Therefore, small changes in temperature have large effects on the amount of energy emitted.

Most important for plants, and for all life on Earth, is the radiant energy emitted by the sun. The thermonuclear reactions in the sun cause the surface of the sun to have an apparent temperature of about 5900 K. Therefore, the radiant energy emitted by the sun is estimated to be about 6.9×10^7 J m^{-2} s^{-1}, which is a huge amount of energy. Sufficient energy is emitted from the sun *every second* to meet the energy use of the Earth's 5.9 billion people at the level of industrialized consumption for roughly 50 thousand years!

The energy emitted by a black body is distributed over a well-defined wavelength region that depends on the surface temperature. For the sun, most of the energy is emitted over a waveband region of 200–3000 nm (Fig. 8.1). The wavebands in the 400–700 nm region are particularly important for plant life. Photons in this region have sufficient energy to trigger photochemical changes in specialized molecules or pigments. The 400–700 nm range waveband is called 'photosynthetically active radiation' (PAR). The photons in this waveband are absorbed by chlorophyll to support photosynthesis. A small portion of the sun's energy is emitted in the highly energetic ultraviolet wavebands (<400 nm). The remaining wavebands in the 700–3000 nm range contain almost half the solar energy and are called the solar infrared wavelengths. The energy in the photons of the solar infrared waveband is generally too low to trigger photochemical activity.

The wavelength of maximum energy flux density emitted from a black body is also dependent on the temperature of the black body. For the sun, with an approximate surface temperature of 5900 K, the wavelength of maximum photon flux density is 490 nm or the color green. Because the maximum emis-

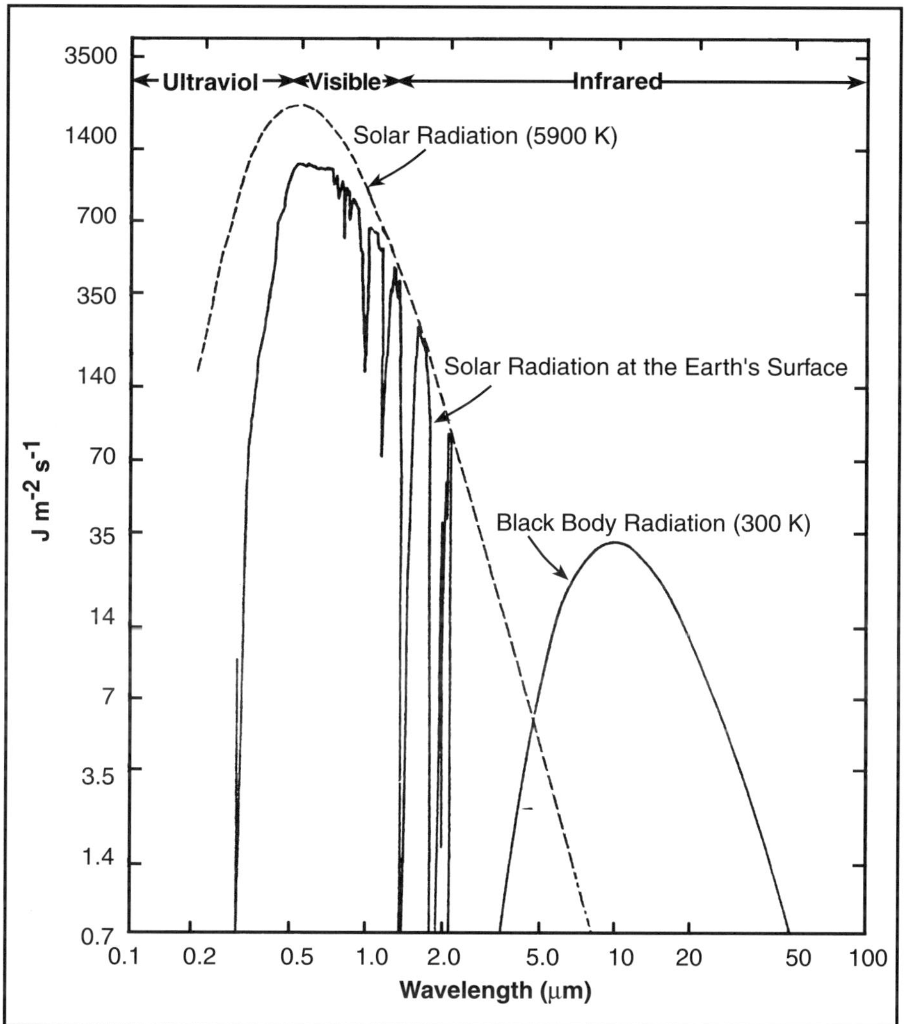

Figure 8.1 ***Radiation energy per unit wavelength plotted against wavelength. The left-hand curve is for solar radiation and the right-hand curve is for surfaces at 27°C (300 K)*** *(redrawn from Sellers, 1965).*

sion of solar energy is at 490 nm, it is not surprising that many biological processes have evolved photochemistry to use this portion of the radiant energy spectrum.

The fraction of the sun's emitted energy that reaches the Earth is very small because of the distance of the Earth from the sun (1.55×10^8 km). The flux density decreases as a function of the square of the ratio between the sun's radius and the radius of the Earth's orbit around the sun [$(6.7 \times 10^5 / 1.5 \times 10^8)^2$]. Therefore, the flux density of solar energy reaching the top of the Earth's atmosphere is calculated to be 1380 J m^{-2} s^{-1}. Actually, measurements on satellites give the mean value as 1373 J m^{-2} s^{-1}. This value is sometimes called the solar constant.

Solar Energy at the Earth's Surface

Two factors greatly influence the irradiance of solar radiation at the Earth's surface: the surface–sun geometry and atmospheric transmissivity.

Geometry

Positioning of the sun directly overhead (solar angle of 90° from horizontal) is required to avoid spreading the radiation over a surface tilted with respect to the incident beams of radiation (Fig. 8.2). Geometry can be used to show that the maximum energy at the Earth's surface is the solar constant multiplied by the sine of the solar angle. Therefore, if the solar angle is 60° then the maximum energy received on the Earth's surface is $1373 \times \sin(60) = 1190$ J m^{-2} s^{-1}.

The daily rotation of the Earth and the orbit of the Earth around the sun cause the solar angle to be continually changing at any single position on the Earth. At sunrise and sunset the solar angle is 0° and reaches its maximum at solar noon. The maximum solar angle is dependent on the latitude on the Earth and the position of the Earth in its orbit around the sun. Between latitude 23.5° N (Tropic of Cancer) and 23.5° S (Tropic of Capricorn) the maximum solar angle at midday is 90° at any single location on 2 days each year. At the equator the 90° solar angle at midday occurs on about 21 March and 21 September (*i.e.*, the vernal and autumnal equinox, respectively). The midday solar angle is at 90° for only a single day at 23.5° N (about 21 June) and at 23.5° S (about 21 December). At latitudes greater than 23.5°, the maximum solar angle is always less than 90°.

Atmospheric transmissivity

Atmospheric transmissivity has a very large effect on the quantity of solar radiation that actually reaches the Earth's surface. The atmosphere is only semi-transparent to solar radiation because of reflectance and absorptance by the gases and suspended particles in the atmosphere. Light scattering caused by the gas molecules or large particles causes substantial decreases in the amount of radiation reaching the Earth. For a clear atmosphere, about 30–40% of the radiation incident to the atmosphere will not reach the Earth's surface. Therefore, it is possible to calculate the amount of radiation at the Earth's surface by combining the atmospheric effects with the geometrical effects of a 60° solar angle. The maximal radiation at the Earth's surface will be about

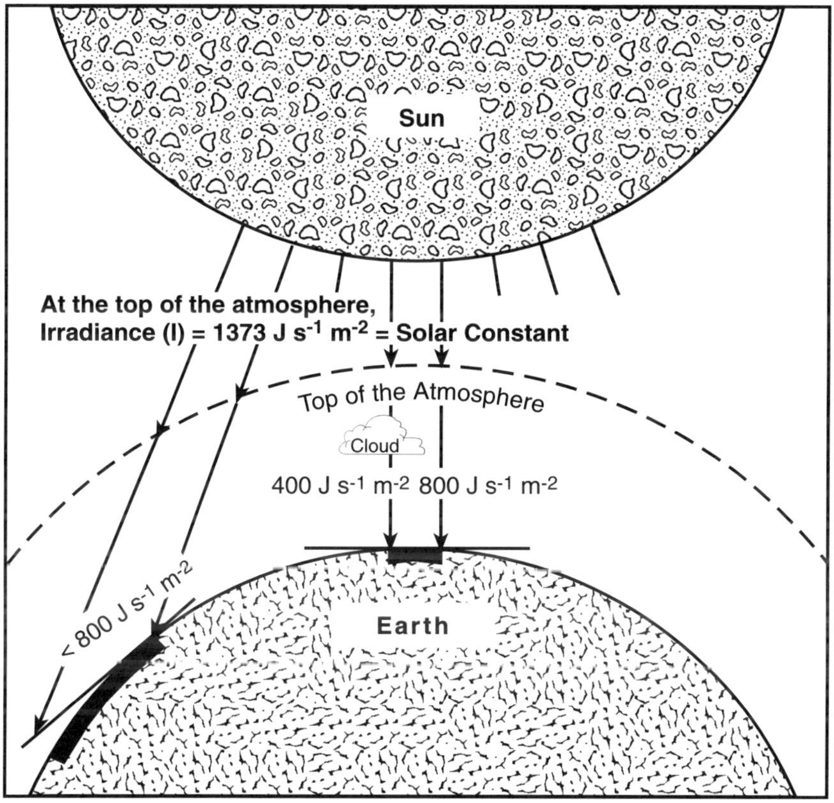

Figure 8.2 ***Schematic of the pathway of solar radiation from the sun to the Earth's surface.***

710–830 J m^{-2} s^{-1} at midday during the summer for many mid-latitude agricultural regions.

The loss of radiation caused by water in the atmosphere is striking, particularly when the water is aggregated into fog or clouds. 'Hazy' conditions that can be attributed to high humidity, suspended particles, or some gaseous pollutants also decrease the amount of radiation reaching the Earth's surface. Because the radiation pathlength in the atmosphere is important in determining the amount of radiation scattering, at low solar angles atmospheric effects on solar radiation losses are greatest.

The scattering of solar radiation in the atmosphere has important consequences other than a loss in radiation. Atmospheric scattering of radiation means that when the radiation reaches the Earth's surface, it appears to originate from angles in the sky other than directly from the sun. This sky radiation is the diffuse component of radiation at the Earth's surface. Diffuse radiation is actually more efficiently intercepted and used by plant canopies than direct beam radiation. Consequently, the decrease in photosynthetic activity because of atmospheric scattering is somewhat less than anticipated solely from the decrease in radiation.

The atmospheric radiation scattering also has important influences on the quality or color of the radiation reaching the Earth's surface. Because the shorter wavelengths (*i.e.*, blue color of the visible spectrum) are scattered to a greater degree than longer wavelengths, the sky under clear conditions appears blue. At lower sun angles especially, the loss of shorter wavelengths from the direct beam radiation causes the sun to appear orange or red. Importantly for plants, at sunrise and sunset the relative levels of various wavelengths are changed. Because the shorter wavelengths have been scattered by the atmosphere at this time of day, plants are subjected to higher levels of red than blue at midday. The exposure to relatively high levels of red wavelengths may be important in triggering photosensitive processes in plants.

Surface irradiance

The combined effects of surface–sun geometry and atmospheric transmissivity explain much of the geographical variation of solar radiation on the Earth's surface (Fig. 8.3). Near the poles, because of low solar input during winter months, annual receipt of solar radiation is only 20–25% of that in the tropics. The highest annual input of solar energy is at the Tropics of Cancer and Capricorn rather than at the equator. This is because the climate at the equator is generally cloudy and rainy compared with higher latitudes. In fact, deserts commonly exist at the latitudes of the Tropics of Cancer and Capricorn because of the lack of clouds and humidity. Consequently, the highest solar radiation receipt is in the Sahara Desert (Tropic of Cancer). These high solar inputs cause a highly positive yearly energy balance and potentially high plant productivity, when water is made available.

Figure 8.3 presents a geographical plot of solar energy in the USA for the months of July and January. Solar radiation for January in northern USA (Fig. 8.3B) is about 4–8 MJ m^{-2} day^{-1}, while it is 23–25 MJ m^{-2} day^{-1} for July (Fig. 8.3A). Latitude has little effect. Atmospheric turbidity (smoke, humidity, and dust) does influence radiation levels. The July values in desert climates are about 30 MJ m^{-2} day^{-1} as compared with about 22 MJ m^{-2} day^{-1} for the Great Lakes region.

Most important in agriculture is the fact that the solar input during the growing season throughout the northern hemisphere is fairly stable, except for variations resulting from atmospheric turbidity. As discussed in Chapter 5, dry matter yield from photosynthesis is directly related to the amount of solar radiation intercepted by the crop canopy. Consequently, crop production in May, June, and July are potentially about the same throughout the northern hemisphere regardless of latitude (Fig. 8.3). However, atmospheric conditions may vary widely due to turbidity, causing variations in irradiance at the Earth's sur-

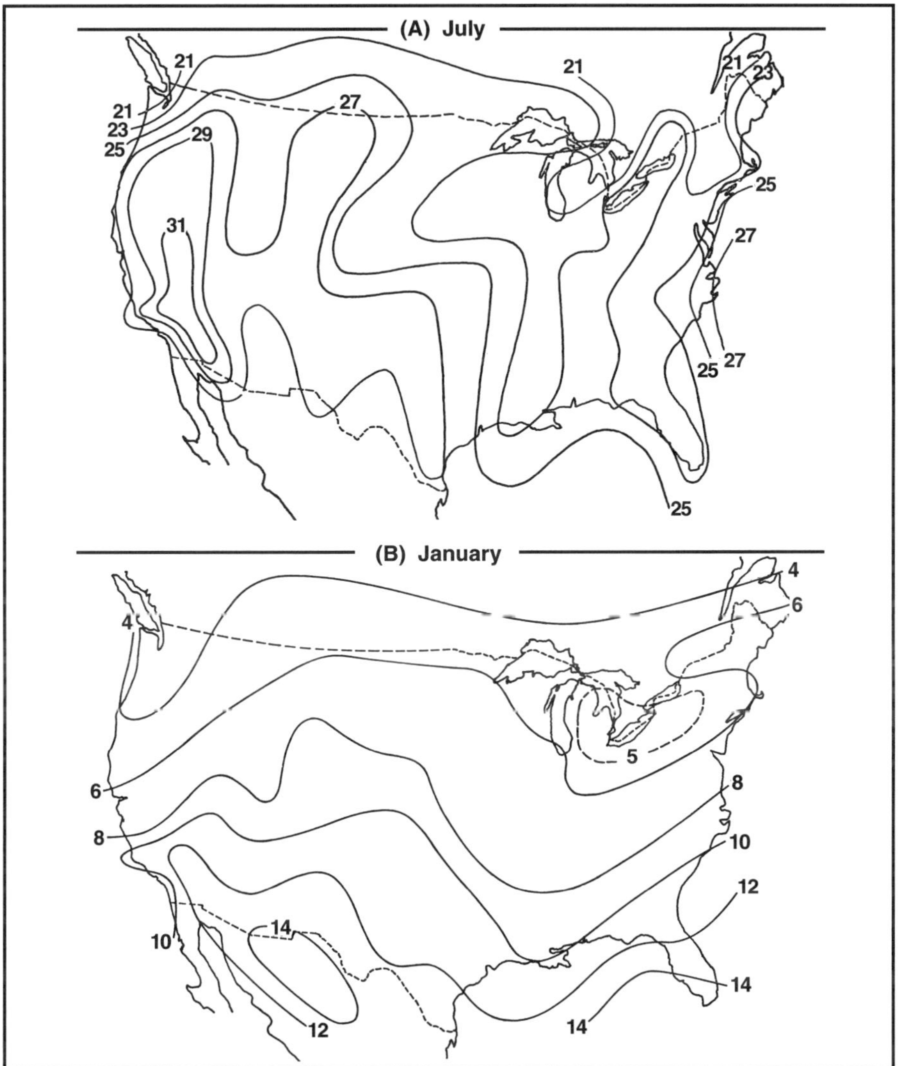

Figure 8.3 ***Average solar radiation (MJ m^{-2} day^{-1}) in the United States in (A) July and (B) January*** *(redrawn from Sellers, 1965).*

face. Desert-like climates, such as at Davis, California, receive an average of about 25 MJ m^{-2} day^{-1} or more of solar radiation during the growing season (May through September). As a result, crop yielding potentials at Davis are about 15–25% more than at a similar latitude with a humid climate, such as in central Missouri.

The fact that biomass yield potential is fixed by solar radiation makes reports of phenomenally high yields highly suspicious. While very high crop yields can be obtained by high applications of mineral and organic fertilizers plus irrigation, solar radiation still sets the ultimate limit. Comparatively lower crop yields occur at the equator, due to persisting heavy cloud cover and high humidity that may severely decrease solar input. Beyond high latitudes, yields are relatively low due to shortness of growing season, early-maturing genotypes, and low temperatures. High yields from cool-season crops (cereal grains and cole crops) in Europe are exceptional due to mild winters allowing for long growing seasons.

Physiological Responses to Solar Radiation

Photosynthesis

The absorption of solar radiation by chlorophyll to provide energy for carbon dioxide fixation is the fundamental process supporting life on Earth. The processes of interception, absorption, and formation of chemical from solar energy were discussed in Chapter 5. It is difficult to overemphasize the importance of photosynthesis and the formation of chemical energy in natural and plant production ecosystems. The goal of most plant production ecosystems is to establish conditions for high rates of photosynthesis in plants.

Photoperiodism

Length of day (photoperiod) or length of night per 24-h cycle is an important signal for many morphogenic responses in plants, *e.g.*, stem elongation, flowering, and fruiting. Photoperiodic responses require only low levels of illumination, *i.e.*, the presence or absence of light, and are not directly related to photosynthesis. The discovery of photoperiodism is credited to Garner and Allard (1920, 1923), initially working with two short-day species, tobacco cv. 'Maryland Mammoth' and soybean cv. 'Biloxi'. The tobacco did not produce flowers in the field at Arlington, Virginia (38° N) before frost, but flowered readily in the greenhouse during the short days of fall. The daylength had to decrease to a critical level before floral initiation and flowering occurred. Biloxi soybean flowered during the short days of autumn in the field regardless of the date of seeding. They concluded that because tobacco and soybean flowering were promoted by the relatively short daylengths of fall, they are short-day plants. Another group of plants (wheat, clovers, beets, and temperate grasses, for example, orchard grass, timothy) flowered in the long days of spring and were classified as long-day plants. Still other plants, for example, buckwheat, tomato, were insensitive to daylength and were classified as day-neutral plants.

Subsequently, Garner and Allard (1923) discovered that length of night, or nyctoperiod, rather than length of day (photoperiod), was the causative factor. Interruption by a few minutes of low-intensity light, destroyed the long-night effect. As a result, plants behaved as if the nights were short and the days were long, *i.e.*, short-day plants remained vegetative and long-day plants flowered.

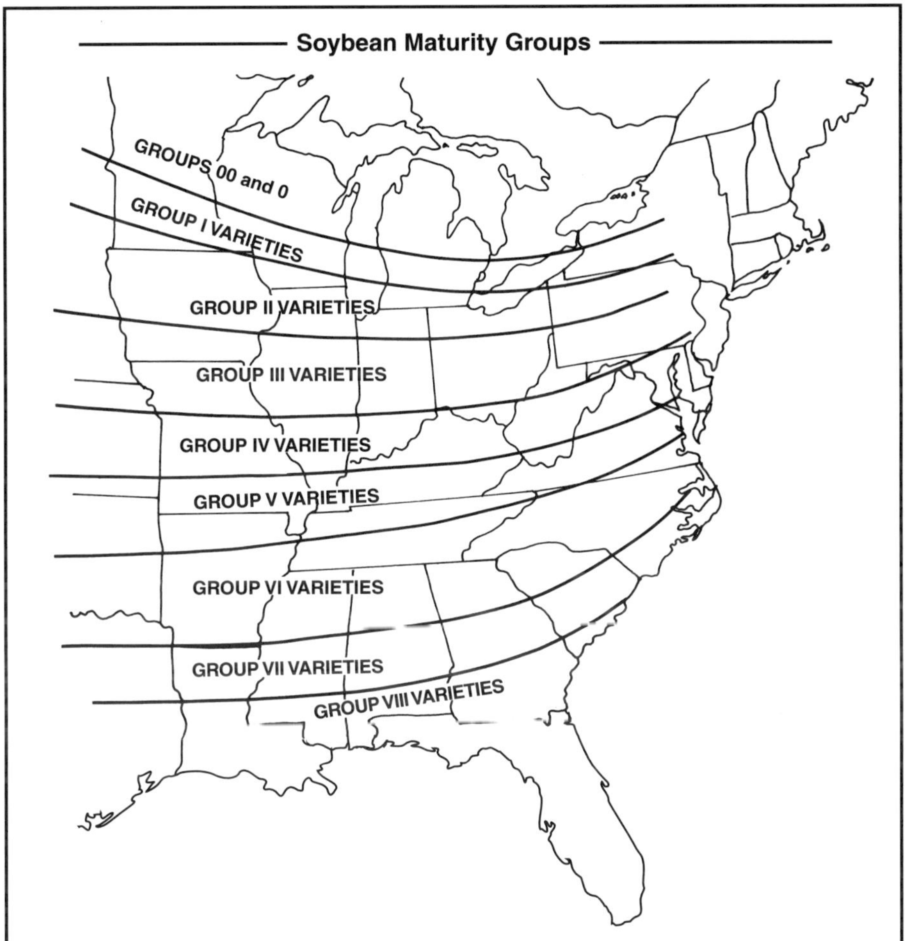

Figure 8.4 ***Zones of adaptation in the United States for various maturity groups of soybean*** *(Poehlman, 1979).*

Only a 2-minute interruption was required to cause the effect in highly sensitive plants such as cocklebur. On the other hand, interruption of the long day by a period of darkness had no effect. Clearly the length of the night period was the critical factor in the 24-h cycle response.

Day-neutral plants mature normally over a wide range of latitudes, whereas those with long-day and short-day plant adaptations are restricted to specific latitudes. Cultivars of soybean, a short-day plant, have well-defined maturity groups and can be successfully produced only in a north–south range of about 300 km (Fig. 8.4). Poleward movement from optimum latitude for an individual cultivar results in excessive vegetative growth. This results because summer days of high latitudes are longer than daylengths in the area of adaptation. Frost may occur before maturation. On the other hand, moving a short-day plant toward the equator results in a reduction of node numbers, leaf size, internode length, and yield. A considerable part of the success in the worldwide distribution of high-yielding 'Green Revolution' crop varieties was a result of a minimization or elimination of photoperiod sensitivity in high-yielding varieties.

Ultraviolet radiation

A small fraction of the solar radiation reaching the top of the atmosphere has wavelengths shorter than the visible wavebands (<400 nm). Wavelengths shorter than blue light in the range less than 400 nm have been labeled as ultraviolet radiation. This region has been further broken down into subregions identified as UV-A (320–400 nm) and UV-B (280–320 nm). Normally, the ozone in the stratosphere of the atmosphere absorbs much of the ultraviolet radiation, especially UV-B, before it reaches the Earth's surface.

A general concern for the effects of ultraviolet radiation on plant growth has developed recently because of ozone layer depletions measured over Antarctica. The energy per photon of radiation is especially high for ultraviolet radiation. The energy of UV-B radiation is sufficiently high to break many biochemical bonds including those in nucleic acids, resulting in an alteration, or even destruction, of molecules. So, projected increases in the amount of UV-B radiation intercepted by plants has been hypothesized to inhibit or damage plant growth and development.

Experiments to observe the effects of UV-B radiation on plants are difficult because of the complexity in artificially varying the flux density of UV-B radiation. The results from several studies have shown that UV-B can have substantial influences on plant development and growth. Unfortunately, many of these studies have compared plant performance between exposure to no UV-B and plants with unrealistically high exposures to UV-B. These experiments are not appropriate for defining the effects of a realistic enhancement in UV-B exposure. Many of these experiments are also of limited use because they were done under conditions of low PAR, which tend to magnify the effects of UV-B radiation.

A few experiments have been done under field conditions where the comparison has been between current exposures to UV-B radiation and to enhanced exposures to UV-B. Nearly all these experiments using enhanced levels of UV-B have shown no or only marginal effects of the increased UV-B levels on plant development and growth. In some experiments, an individual variety might show some effects of the enhanced UV-B treatment but the bulk of the varieties showed no effects.

It appears that plants have generally evolved adequate defense mechanisms to UV-B exposure. The cuticle that covers plant tissue has been found an especially effective absorber of UV-B radiation so that the radiation fails to even enter plant cells. In addition, the stimulation of the synthesis of UV-B absorbing flavonoids under UV-B treatment helps to protect plant cells.

Overall, it appears that projected increases in UV-B radiation are not likely to have any major deleterious effects on plant development and growth. While selected cultivars may be adversely affected, it appears the current range of germplasm is adequate to improve the genetic background of these sensitive cultivars.

Radiational energy balance

Though leaves absorb little of the solar infrared radiation, considerable energy is retained by leaves through the absorption of the PAR wavelengths. While some absorbed PAR energy is stored as photochemical energy in plant pigments, these pigments have only discrete energy levels that are photochemically stable. The excess absorbed energy is dissipated as heat. Further, the biochemical reactions leading to the synthesis of plant constituents result in loss of energy as heat. Leaves will ultimately retain in the biochemical constituents of the plant only about 10 J m^{-2} s^{-1} of the roughly 350 J m^{-2} s^{-1} absorbed solar energy. This means that only about 3% of the absorbed solar energy at midday on a clear day is fixed into chemical energy. To avoid intense heating that might occur from the accumulation of excess heat energy, leaves need efficient means to dissipate heat.

The greatest amount of heat is lost by plants through the process of 'black body' radiation (Fig. 8.1). As discussed previously all surfaces radiate energy in proportion to the fourth power of their surface temperature. For leaves at approximately 27°C (300 K), the radiant energy loss is calculated to be roughly 460 J m^{-2} s^{-1} for each side of the leaf.

The wavelengths at which the radiant energy is lost by plants is also dependent on surface temperature. Again assuming a leaf at 27°C, the wavelength of maximum loss is 9700 nm or 9.7 μm (Fig. 8.1). These wavelengths are in the infrared zone but they are much longer than those originating from the sun. They are called the thermal infrared wavelengths.

The above calculations of solar radiation absorption (350 J m^{-2} s^{-1}) and radiant heat loss from leaves (2 × 460 J m^{-2} s^{-1} in the above example) show a net loss of energy from leaves. This situation would result in rapid freezing unless additional energy is absorbed by leaves. This situation is averted because leaves also readily intercept thermal infrared radiation emitted from surfaces around them. For a single horizontal leaf, emitted thermal infrared radiation is received from surfaces below it including other leaves and soil, and from the atmosphere above it (Fig. 8.5). Assuming the surfaces below the leaf are at temperatures roughly equivalent to the leaf of interest, then the radiant energy intercepted on the bottom surface matches that emitted.

The top surface of the leaf 'sees' a sky that has an apparent temperature substantially below leaf temperature. A dry, cloudless sky may have an apparent temperature of about −53°C (220 K). Therefore, the thermal radiation received by the top side of a horizontal leaf is about 130 J m^{-2} s^{-1}. On balance, to stabilize temperature the leaf only needs to lose a small amount of energy (350 − 460 + 130 = 20 J m^{-2} s^{-1}) by other mechanisms. On the other

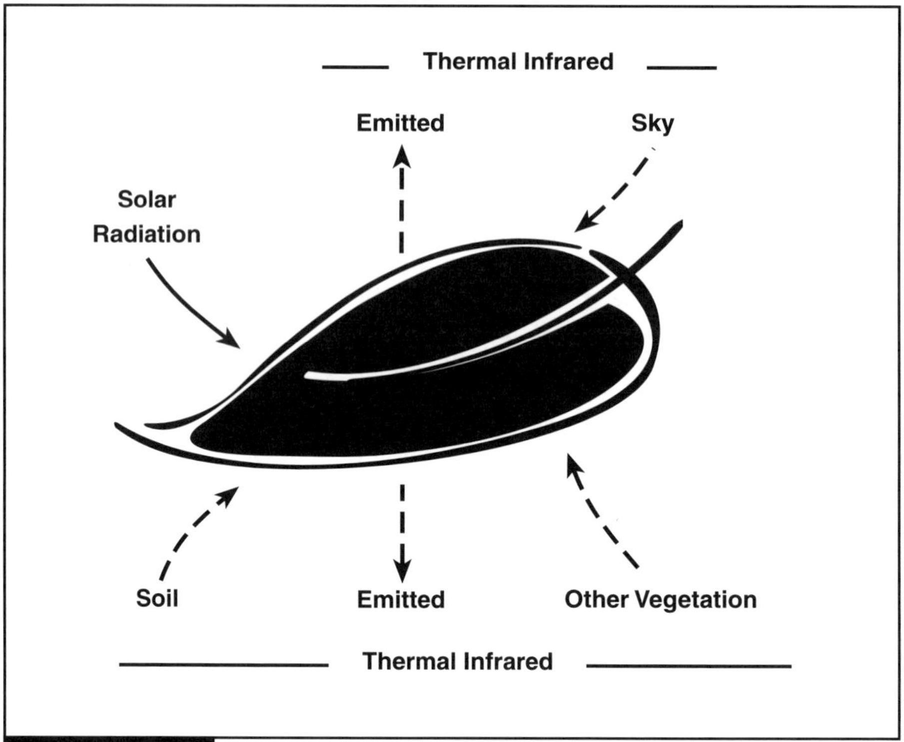

Figure 8.5 ***Schematic of the radiation transfer to and from an individual leaf.***

hand, a cloudy sky has a considerably higher apparent temperature that may be as high as 7°C (280 K). Calculation of the thermal radiation emitted by a cloudy sky gives an estimate of about 350 J m^{-2} s^{-1}. Because the amount of incident solar radiation may also be decreased by half under a cloudy sky, the energy budget again shows a small positive balance (350/2 − 460 + 350 = 65 J m^{-2} s^{-1}). Again, other mechanisms are needed to dissipate the excess heat.

The above calculations help to illustrate the paradoxical fact that the radiant energy balance of a leaf can be particularly important at night! Deleting the energy input of solar radiation in each of the above calculations, the net radiant energy budget for each case is negative (0 − 460 + 130 = −330 J m^{-2} s^{-1} under a cloudless sky; 0 − 460 + 350 = −110 J m^{-2} s^{-1} under a cloudy sky). Not surprisingly then, leaf surfaces as well as most surfaces at the Earth's surface, cool at night because of this negative radiant energy balance. Obviously, sky conditions greatly affect the rate and extent of cooling at night. On clear nights, the temperature of leaves can actually cool to temperatures below freezing although the air temperature is above freezing. This phenomenon results in 'radiational frost'. High value crops can be placed under artificial coverings during these clear sky conditions so that the 'sky' temperature is that of the cover, or roughly the air temperature. The simple procedure of covering plants can greatly increase the apparent 'sky' temperature for the plants. That is, from the calculations above the radiant energy budget is moved from approximately $-330\ \text{J m}^{-2}\ \text{s}^{-1}$ to $-110\ \text{J m}^{-2}\ \text{s}^{-1}$. Slowing the radiant energy loss by one-third may prevent frost and freezing of plants during the night under many circumstances.

Another issue involving the radiant energy balance is the 'greenhouse' phenomenon discussed in detail in Chapter 11. Increasing concentrations of certain gases such as carbon dioxide in the atmosphere has the effect of increasing the apparent sky temperature. Even a very small increase in apparent sky temperature when taken to the fourth power, can result in a slight increase in the radiant energy balance at the Earth's surface. This results in what has been commonly called 'greenhouse' heating. To a large extent, climate heating predicted from increasing atmospheric concentrations of 'greenhouse gases' is a consequence of an anticipated change in the sky radiant energy input.

Summary

All life requires energy input to survive and to sustain the integrity of the organism. The ultimate source of most of the energy consumed by humans is the sun through chemical energy conservation in plant tissues. Fossil fuels are derived from plant tissues that stored solar energy many millions of years ago. The amount and nature of the energy released by the sun are well defined by the laws describing radiation physics. Accounting for the transport of the energy to the orbit of the Earth, the transmission of the radiation through the Earth's atmosphere, and the geometry of incident radiation at the Earth's surface results in a good estimate of the radiant energy available to plants. Differences in atmospheric conditions that influence radiation transmission through the atmosphere, account for much of the geographical variation in radiation.

Several physiological processes are directly dependent on the interception of solar radiation. Most important, radiant energy must be absorbed by leaves to support the photosynthetic fixation of CO_2. The photoperiod is also very important in influencing developmental processes in plants. There are short-day, long-day, and day-neutral species and these response traits are very important in defining the latitudinal zones of acclimation for individual species and cultivars.

Including the thermal infrared wavebands in the energy budget of plants helps to explain the deviations in temperature experienced by plants from air temperature. During the day, the radiation balance of leaves is generally positive so that other mechanisms may be required to dissipate the extra heat absorbed by leaves. At night the radiation balance is negative and explains why leaf temperatures go below air temperature, and with cool air temperatures plants can be exposed to radiational frost.

Further Reading

Cox, G.W. and Atkins, M.D. (1979) *Agricultural Ecology*. W.H. Freeman, San Francisco.

Garner, W.W. and Allard, H.A. (1920) Effect of the relative length of day and night and other factors of the environment on growth and reproduction in plants. *Journal of Agricultural Research* 18, 553–607.

Garner, W.W. and Allard, H.A. (1923) Further studies in photoperiodism, the response of the plant to relative length of day and night. *Journal of Agricultural Research* 23, 871–921.

Hills, F.J., Johnson, S.S., Abshahi, S.G.A. and Peterson, G.R. (1981) Comparison of high-energy crops for alcohol production. *California Agriculture* 35, 14–16.

Hopkinson, C.S., and Day, J.W. (1980) Net energy analyses of alcohol production from sugarcane. *Science* 207, 302–303.

Monteith, J.L. and Unsworth, M.H. (1990) *Principles of Environmental Physics*. Edward Arnold, London.

Pimentel, D. and Hall, C.W. (1984) *Food and Energy Resources*. Academic Press, Orlando, Florida.

Poehlman, J.M. (1979) *Breeding Field Crops*. AVI Publishing, Westport, Connecticut.

Prine, G.M., Dunavin, L.S., Brecke, B.J., Stanley, R.L., Mislevy, P., Kalmbacher, R.S. and Hensel, D.R. (1988) Model crop systems: Sorghum, napiergrass. In: Smith, W.H. and Frank, J.R. (eds) *Methane from Biomass, a Systems Approach*, Elsevier Applied Science, Essex, pp. 83–102.

Ramade, F. (1984) *Ecology of Natural Resources*. John Wiley and Sons, New York.

Sellers, W.D. (1965) *Physical Climatology*. University of Chicago Press, Chicago.

Terry, N. (1979) Photosynthesis and plant development. In: Marcelle, R., Clijsters, H. and Van Poucke, M. (eds) *Proceedings of Symposium, Diepenbeck-Hasselt, Belgium, July 1978*, Junk, the Hague, pp. 151–160.

Temperature

9

K.J. BOOTE AND F.P. GARDNER

Temperature controls the rate of physiological processes and sets the limits of plant survival. Consequently, it is overriding in importance as an ecological factor in determining the adaptation and distribution of crop and native plant species. The temperature range of physiological processes is narrow (5–40°C), including photosynthesis and growth. The cardinal temperature points, minimum, optimum, and maximum, for growth of a species define this range. Contrasted with growth, many perennials can survive temperatures ranging from −50 to 50°C. Ephemeral species can avoid temperature extremes by maturing seeds during the growing season, and allow only the seeds to survive the temperature extreme. Temperature, however, limits plant production of certain species to the tropics (0° latitude ± 23.5° N,S). Bacteria may survive wide extremes, ranging from 100°C to near −270°C.

Global Temperatures

Temperature varies according to season, latitude, altitude, proximity to bodies of water, and local factors. The southern hemisphere is 81% ocean compared with 61% ocean for the northern hemisphere. Also, it is seen that except for Antarctica, the northern hemisphere also has more high mountains (>3000 m). Temperature decreases with increasing altitude at 5–6°C per 1000 m. The result overall is that the northern hemisphere is colder than the southern hemisphere and other climatic parameters are affected.

Temperature also decreases with increasing latitude due to less solar energy input. Therefore, increases in altitude and/or latitude result in a shift toward ecosystems adapted to survive cold temperatures, such as the tundra. Reported air temperature extremes range from 58°C (136°F) at Azizia, Libya, to −71°C (−96°F) at a northern Siberian weather station. The record low of −86°C (−122°F) was recorded at a Russian station in Antarctica, at an extremely high latitude and high altitude location. The highest mean annual temperature of

30°C (86°F) was recorded at Massawa, Eritrea (Africa). The lowest mean of −26°C (−14°F) was recorded at Framheim, Antarctica.

Global mean temperatures for January and July are shown in Fig. 9.1. Summer highs are greatest in the Sahara, American, and Asian deserts near the Tropic of Cancer. Large bodies of water produce marine climates, which moderate temperature.

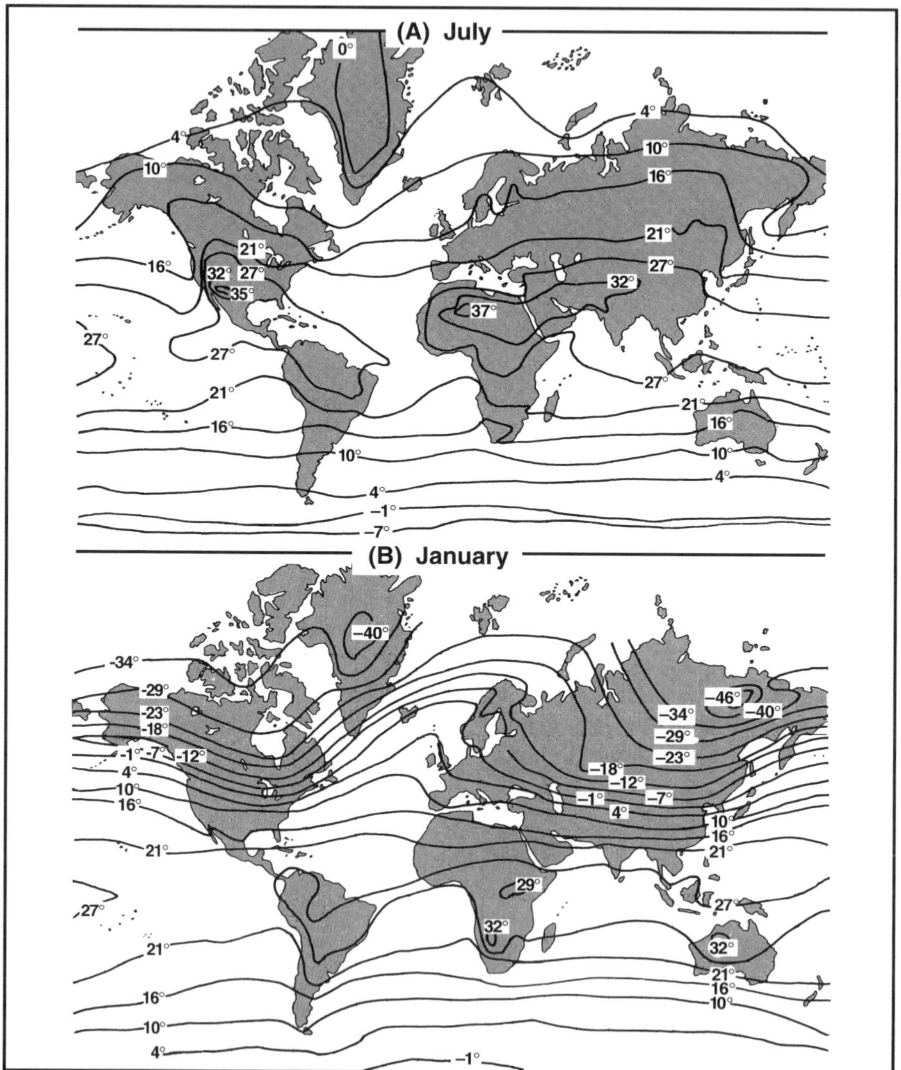

Figure 9.1 ***Average monthly temperatures worldwide in (A) July and (B) January*** *(redrawn from Miller and Thompson, 1979).*

Continental Effects

The extent of seasonal and diurnal temperature variation provides an index of continental effects. Continental extremes, such as those which occur in central Asia and North America, may vary by as much as 65°C (149°F). High latitudes (>40°) have negative energy balance in winter (loss exceeds solar energy input). Large water bodies buffer temperature against changes because the heat holding capacity (specific heat) of water is high. In contrast with the northern hemisphere, most of the southern hemisphere is covered by ocean, so temperatures at high southern latitudes are moderate. With its greater land mass, the northern hemisphere has greater occurrence of cold-air masses formed in polar areas in winter. These air masses move toward the tropics, often producing severe freezes in the northern hemisphere.

Seasonal and Diurnal Temperature

The energy source for increasing temperature is solar radiation, which peaks daily in intensity at solar noon and peaks annually in the northern hemisphere on 21 or 22 June (summer solstice). The daily low temperature is at about 0700 h or at sunrise, whereas the high is at about 1500 h. Daily temperature peaks lag behind the daily peaks of solar radiation by 3 h. During summer, the daily maximum and minimum temperature in humid temperate climates may differ by 10–12°C. Diurnal variation is much greater than 12°C in desert climates, perhaps as much as 30°C. Seasonal temperature peaks lag behind the peaks in solar radiation by 1 month. In the northern hemisphere, seasonal temperature highs occur in July and lows in January.

Energy Transfer Processes

Energy is transferred from one object or medium to another as (i) *sensible heat*; (ii) *latent heat*; and (iii) *radiation*. All of these transfer processes are important in cooling plants and the environment, but the rate depends on thermal properties. The loss (or gain) of energy from objects by thermal infrared radiation is described in Chapter 8. A description of the other two transfer processes follows.

Sensible heat

Sensible heat is the energy gain or loss from molecules by transfer of kinetic energy (molecular motion) to adjacent molecules. Transfer of kinetic energy (heat) in fixed media is called conduction. If the medium is in fluid motion, such as occurs in water or air, the energy transfer is called convection. Convective movements of air masses quickly become a very important aspect of sensible heat exchange between soil, plants, and their environment.

Rate of sensible heat transfer depends on the difference in temperature between surfaces and media. Sensible heat transfer is negative (away from the object) if the object is warm and losing heat to a cooler environment. An object can gain heat if it is cooler than the surrounding environment. Rate of transfer is also highly dependent on the thermal properties of the fluid; air being the most important fluid of convection for plants.

Transfer of heat can also occur through direct molecular contact, consequently, conductivity is directly related to the density of a substance (closeness of molecules). Gold, silver, copper, lead, and water are excellent conductors of heat. Therefore, wet soils conduct heat well since water molecules contact soil particles as well as each other. Dry soils contain air-filled voids, and therefore, are not good conductors. The surfaces of dry soils are hot in summer, since the absorbed solar energy is not readily conducted to cooler levels.

Latent heat

In plant ecosystems, the heat input from solar radiation and sensible heat raises the energy level of molecules. Greater molecular energy results in breaking of hydrogen bonds that bind molecules together. As a result, some molecules escape the liquid state of water and become water vapor (Chapter 7). The energy required to cause a change in state (ice $\leftrightarrow$ liquid $\leftrightarrow$ vapor) without a change of temperature is referred to as latent heat. A greater amount of energy is required to transform liquid water to vapor (14.25×10^{-3} J g^{-1} at 100°C and one atmosphere pressure) than to transform solid ice to liquid (1.91×10^{-3} J g^{-1} at 0°C and one atmosphere pressure). Sublimation is the transformation of ice to vapor (16.16×10^{-3} J g^{-1}). Sublimation is common on frozen ground or ice sheets during winter in temperate climates.

Loss of water as vapor from vegetation (transpiration) and from soil (evaporation) is collectively called evapotranspiration. In evaporation, water molecules with higher kinetic energy leave the evaporating surface; consequently the evaporating surface loses energy and is cooled. Evapotranspiration is the dominant energy transfer process in the biosphere and important in the global energy balance.

Soil Temperature

Soil type

Soils differ in thermal properties depending on their color, and on their air and water content. A darker color soil absorbs more solar radiation and as a consequence, heats more rapidly than a lighter colored soil. For example, in Fig. 9.2 the difference in the soil surface temperature of the peat and clay soil is partly a result of the darker color of the peat soil.

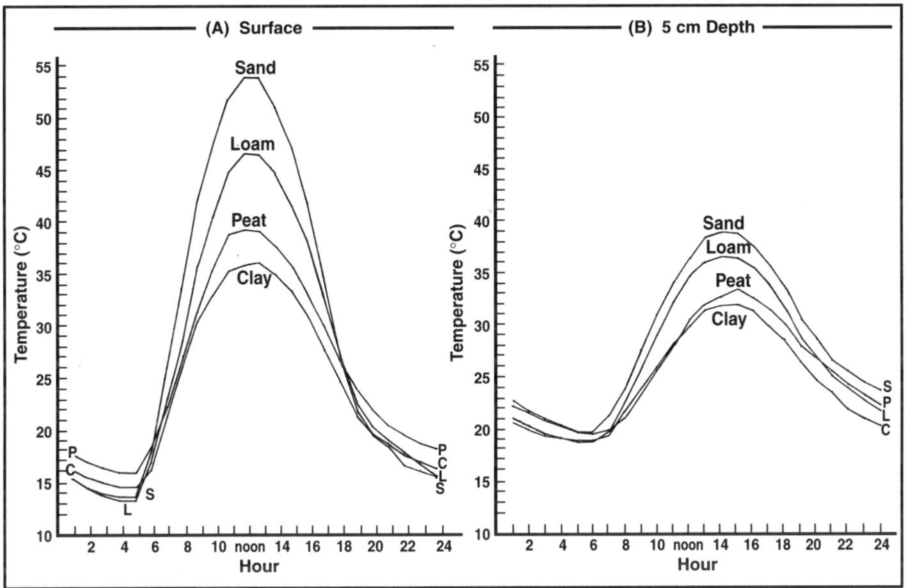

Figure 9.2 ***Soil temperatures in July (A) at the surface and (B) at a 5 cm depth (on a clear summer day in Japan) for four soil types under conditions where the soil surfaces were kept bare*** *(redrawn from Chang, 1968).*

The texture of the soil is also related to the temperature increase of soils. As illustrated in Fig. 9.2, the greatest temperature increase of the soil surface is for sandy soil followed by loam and then by clay soil. This difference in heating is not directly related to the soil texture *per se*, but rather to the differences in soil water content associated with the different soil textures. Sands hold small amounts of water and have large volumes of airspace. The air in the sand has a small heat capacity and a low thermal diffusivity. The soil surface is thermally isolated and the temperature increases rapidly. This is readily confirmed by anyone who has walked barefooted on the dry part of a sandy beach on a bright day.

In contrast, clay soils hold a large amount of water so that their heat capacity and thermal diffusivity are high. Thus, for clay soils more energy is required to heat the surface layers than for lighter textured soils. Also, the heat accumulated at the surface of a clay soil is more rapidly conducted downward in the soil.

Organic soils also tend to heat more slowly for the same reason as clay soils (Fig. 9.2). The organic matter in the soil has a high water holding capacity that imparts a high heat capacity to the soil. Therefore, high organic soils tend to be cold in the spring and warm in the fall.

Soil depth effect

Diurnal and seasonal temperature of soil varies with soil depth. Increasing depth affects soil temperatures as follows: (i) the amplitude of fluctuations over time (sine curve) is dampened; and (ii) fluctuation peaks are lagged. Therefore, soil-temperature maxima at increased depths are smaller and they occur later in the day (Fig. 9.3). For example, maximum temperature near the surface (0.01 m), at 0.05 m, and at 0.20 m was at a maximum at about 1400, 1600 and 2000 h, respectively. The diurnal soil temperature at 0.8 m remained constant. Seasonally, soil temperature at 0.03 m normally peaks in July, but at 7.5 m it peaks in December.

Vegetation and mulch effects

Vegetation and mulches affect soil temperature in at least two ways. First, they shield the soil from incoming solar radiation and thereby reduce energy absorption. Second, mulches sustain a higher moisture level that results in a higher thermal capacity near the soil surface. Reduction of net energy by vegetation or by mulch cover varies with canopy density and type or thickness of the mulch cover. Mulches and vegetation absorb solar energy and dampen the amplitude of the sinusoidal-wave curves of soil temperature. Figure 9.3 illustrates clearly how sod cover dampened the diurnal amplitudes of soil temperature compared with bare soil. Mulches act similarly in this respect. Further, increased soil moisture content and soil heat capacity cause a lag in amplitude of diurnal temperatures and the time to reach a specific temperature may be delayed.

Ground cover by the growing crop results in less soil absorption of solar radiation and more latent energy loss through the crop. A crop canopy shading the soil can result in the temperature of moist soils being reduced by 6–8°C in comparison with bare soils. Transpiring vegetation also dries the soil and decreases its heat capacity.

Mulching techniques and no-tillage sowing have become common agricultural practices to prevent erosion, control weeds, and conserve moisture. A disadvantage is that soil temperature is often much cooler under no-tillage and mulching. This results because of the combined effects of greater soil moisture retention and lower solar radiation absorption by soil. Energy diffuses to a greater depth in a wet soil than in dry soil. The occurrence of cold wet soils in spring discourages the practice of mulching and other methods of conservation tillage in temperate climates. In the fall, soils under a mulch are warmer than bare soil. Mulching primarily affects maximum temperature; the effect on minimum temperature is small (Table 9.1). Of course, the mean temperature is affected and growth is generally proportional to the mean temperature.

Plastics are used for mulches but clear vinyl accumulates heat in a similar way to a greenhouse, which may cause temperatures sufficiently high to kill weed seeds and other pests. Clear vinyl covering is used to 'solarize' the soil

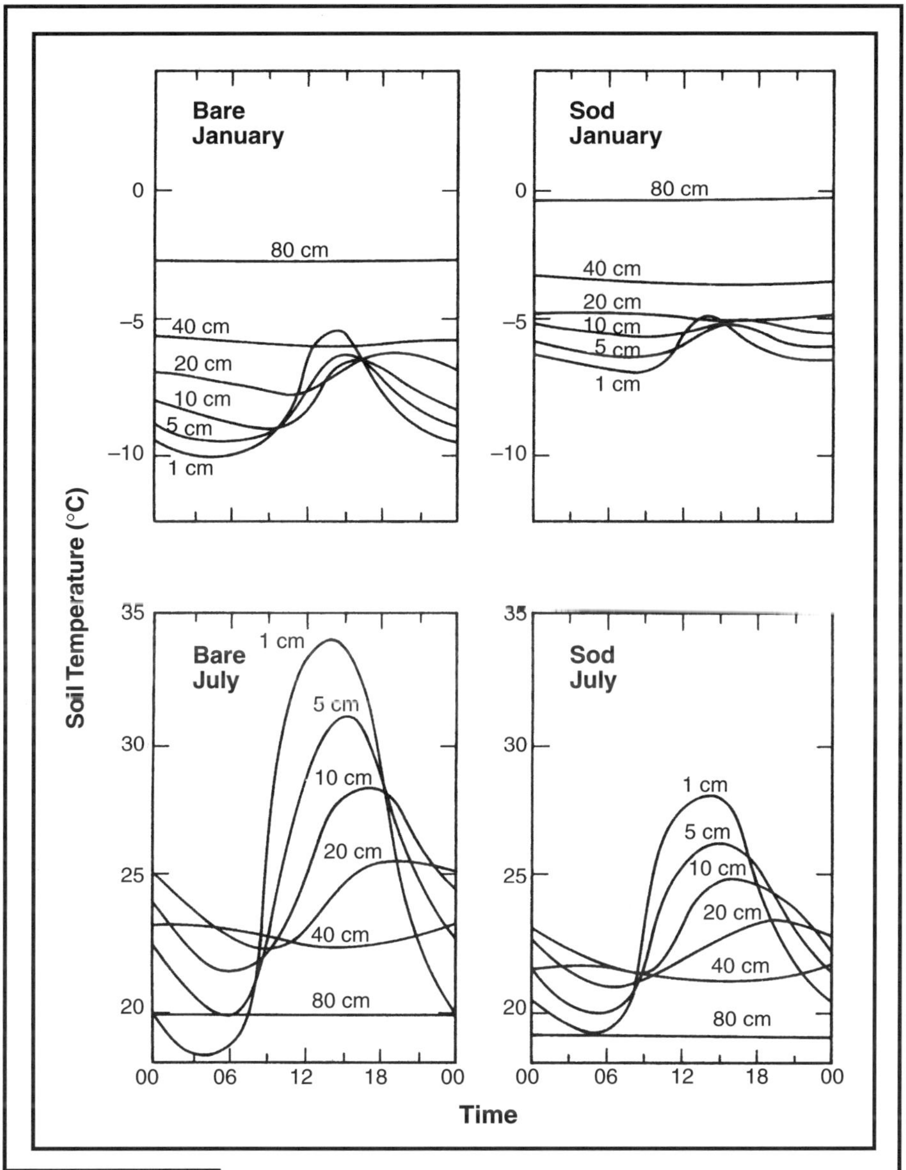

Figure 9.3 ***Soil temperatures at various depths for bare and sodded soil in January and July, at St Paul, Minnesota*** *(redrawn from Rosenberg* et al.*, 1983).*

where solar radiation levels are high, such as in Israel and the Imperial Valley of California. The high soil temperatures that develop under clear vinyl can effectively sanitize the seedbed of weeds and insect pests.

Table 9.1. Effect of vegetation and straw mulches on soil temperature at 2.5 cm depth (McCalla and Dudley, 1946).

	Maximum (°C)	Minimum (°C)
Clean plowed, maize	31.2	17.9
1.8 t straw	26.5	19.6
7.1 t straw	23.6	19.7
Clean plowed, no maize	40.5†	–
Subtilled, 4 t straw	32.7†	–

† Measured on 30 June.

Air Temperature Modification

Effect of large bodies of water

Proximity to large bodies of water such as oceans or lakes can greatly influence annual, seasonal, and diurnal temperatures. The influence of water, however, is primarily linked to prevailing wind direction. The cities of St Louis (continental climate) and San Francisco (marine climate) are at about 40°N latitude. The temperatures in San Francisco remain almost consistently cool year long due to the influence of the cold water currents of the Pacific Ocean. On the other hand, the monthly mean temperature for St Louis has a range of nearly 20°C from January to July and extremes can vary from −20 to 40°C (−22° to 102°F).

Fruit trees on southern exposures or on the windward side of water are warmed faster in the spring and may initiate growth and reproduction. This early growth can be a serious problem because they become susceptible to frost injury. By contrast, winds passing over cold water are cooled and cause a delay in blooming until the frost date has passed. This phenomenon explains the location of many large fruit growing areas, namely the cherry region, east of Lake Michigan. The would-be warm spring winds from the western land area are cooled as they pass over this cold lake. In winter, these westerly winds are warmed by the lake.

Elevation, aspect and slope effects

Elevation and slope orientation (aspect) have pronounced temperature effects. Solar radiation on south exposures in northern latitudes greater than the Topic of Cancer is always from a southward direction. Consequently, the radiation is more direct than that on level land resulting in higher soil temperatures on the south-facing slopes. Accordingly, in the northern hemisphere, summer crops may be produced successfully as high as 2500 m or more on south-facing slopes. This contrasts to maximum altitudes of about 1000 m or less for growing summer crops on north-facing slopes. Farming on a steeper south aspect might allow crop production at an extra 1000 m in elevation and expand the potential arable land area. A north-facing slope would have the same temperature advantage in the southern hemisphere.

Elevation effects on crop selection as a result of temperature differences are particularly noticeable in the tropics. Temperate crops such as potato, bean, wheat and tea are grown at high altitudes (greater than 1500 m). On the other hand, banana and rubber are grown only at relatively low altitudes (less than 1000 m) where temperatures are constantly warm. Some crops, however, have a wide range of adaptation to elevation (e.g. maize, cassava and coffee).

At higher elevations, the air is colder and may be denser than the warmer air at lower altitudes or valleys. The cold air drains down the slope at night, passes under the warm air, and causes a temperature inversion. Shelter belts can be used to reduce frost and cold-air drainage that may damage crops. Cold air can be particularly harmful to plants during reproductive phases.

A warm temperature advantage from aspect can also be obtained by ridging naturally flat ground to create a south-facing slope in the northern hemisphere. That is, the ridges must be constructed to run east to west so that they are normal to the low elevation of the winter sun (Fig. 9.4). Ridging is a common practice in the production of winter vegetables in Florida, but it can be employed at spring planting time on cold, wet soils of higher latitudes. Other

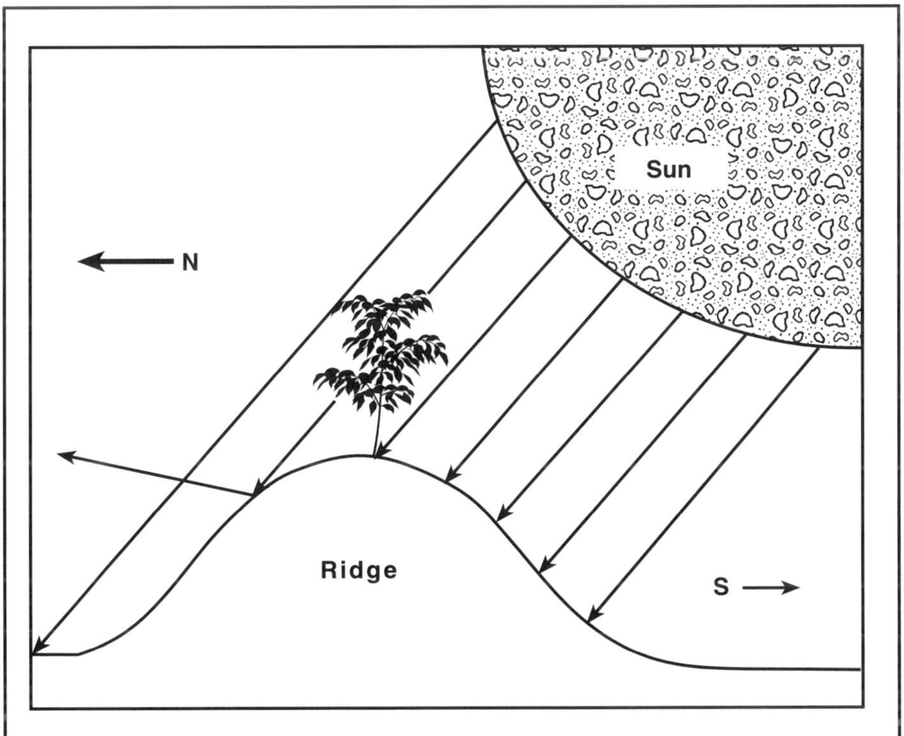

Figure 9.4 *Schematic of rays of solar radiation impinging on surfaces of differing perspective.*

microclimate modifications may include greenhouses, wind machines, proximity to cities, forests, and crop canopies.

Effect of shelter belts

Shelter belts, commonly used in drier continental climates, are used to protect summer crops from hot dry winds and excessive evapotranspiration. Maximum protection is achieved at three tree heights away from the shelter belt. Shelter belts are also used to protect orchards or other crops from cold-air drainage from high altitudes as discussed above. Tall rows of a crop canopy can protect interplanted seedlings from winds but the protection generally is from damage by blowing soil particles.

Plant Development and Temperature

Plant temperature fluctuation, like that of soil, is influenced by the aspect of plant parts. Dense stems and those parts exposed to direct solar radiation have a higher temperature than their shaded or less dense counterparts. Root temperature fluctuation corresponds to that of the soil since roots are essentially a component of the soil continuum.

Temperature profiles in plant canopies fluctuate diurnally. During night and early morning, canopy temperatures are nearly isothermal (same at different canopy depths). At midday, temperatures vary greatly from top to bottom of the canopy. The actual shape of the vertical temperature profile depends on the generation of sensible heat at various levels in the canopy.

Plant development

Plant development rate is highly correlated to temperature. For example, rate of development (e.g., node, leaf, or panicle appearance) is temperature dependent. Node number, plant height, and other development rates are highly correlated with heat accumulation rather than photosynthesis. Figure 9.5 shows that the appearance rate per day for maize leaves increases linearly to an optimum and then declines as temperature exceeds the optimum. Leaf appearance rates of many other species have a similar response to temperature. Although the base temperature used in these calculations varies depending on whether the species is a cool-season or warm-season type.

The rate of progress toward flowering (calculated as the reciprocal of the number of days to flower) or maturity responds to temperature in a similar way to leaf development. The cardinal temperatures for reproductive development, however, are not necessarily the same as those for vegetative development. In several warm-season crops, for example, the optimum temperature for reproductive development is lower than the optimum for vegetative node appear-

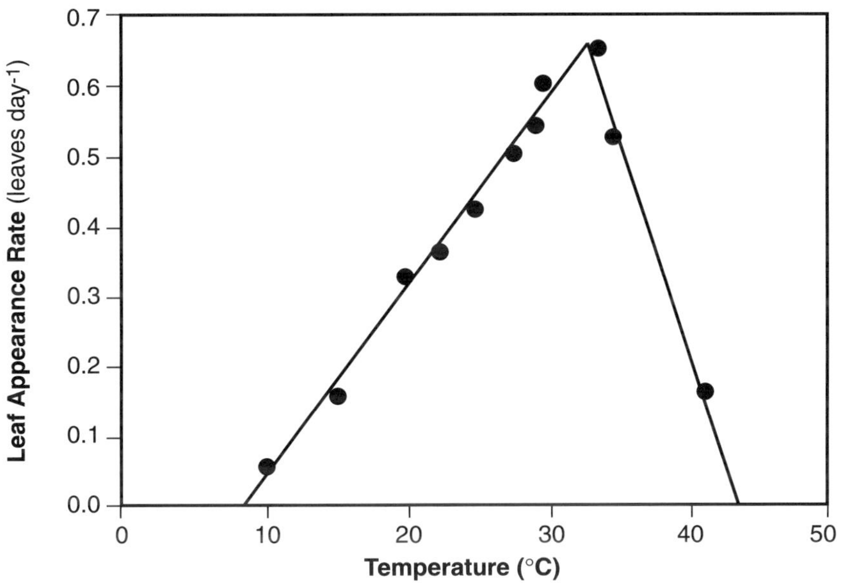

Figure 9.5 ***Appearance rate of maize leaves plotted against temperature*** *(redrawn from Kiniry and Bonhomme, 1991).*

ance. Furthermore, a distinction must be made between progress toward flowering and rate of fruit set. For example, fruit set is limited at temperatures below 18°C in soybean and below 17°C in peanut. Pollen sterility limits grain set in rice when daytime temperature exceeds 35°C.

Chill requirement for flowering (vernalization)

Some species require a period of cold or chilling (2–10°C) to induce flowering, i.e., vernalization. The cold period can generally be calculated as an accumulation of cold units below a base temperature. Species responding to vernalization, usually as growing plants, include winter annuals (wheat, rye, etc.), biennials (sugarbeet, carrot, celery, etc.), and temperate perennial species (orchard grass, bluegrass, apple, etc.). The maximum effective temperature for vernalization of wheat was observed to be 11°C while vernalization did not occur at temperatures as cold as −2°C. For most winter-type wheat cultivars, 6–8 weeks of cold is sufficient to induce heading, but mild winter types respond to only 2–4 weeks of cold. The vernalization time is reduced by plant age and may be zero in older plants, i.e., winter type may eventually flower. Certain temperate perennial grasses required vernalization together with short daylengths as occur naturally in fall to induce flowering in the following spring.

Spring days in high latitudes are long. After sufficient vernalization, long days are generally required to achieve heading or 'bolting' (as in sugarbeet), i.e., plants that respond to vernalization also respond to long days for flowering.

Degree days or thermal units

Maximum, optimum, and minimum temperatures, called the cardinal temperature points, for a wide range of plant species are given in Table 9.2. Most plants are adapted to higher daylight temperatures and cooler night temperatures. If the carbon balance is adequate, developmental events in the life cycle are usually assumed to be determined by mean temperature in the cardinal temperature ranges. With respect to temperature optima, crop species are often placed in two general categories, cool-season and warm-season species. Optimum temperature for the two groups differs by about 10°C.

Table 9.2. Cardinal temperatures for germination of seeds and spores of plant species. (From Larcher, 1980)

	Temperature (°C)		
Plant group	Minimum	Optimum	Maximum
Fungus spores			
Plant pathogens	0–5	15–30	30–40
Most soil fungi	ca. 5	ca. 25	ca. 35
Thermophilic soil fungi	ca. 25	45–55	ca. 60
Grasses			
Meadow grasses	3–4	ca. 25	ca. 30
Temperate-zone grain	2–5	20–25	30–37
Rice	10–12	30–37	40–42
C_4 grasses of tropics and subtropics	10–20	32–40	45–50
Herbaceous dicotyledons			
Plants of tundra and mountains	5–10	20–30	
Meadow herbs	2–5	20–30	35–45
Cultivated plants in the temperature zone	1–3	15–25	30–40
Cultivated plants in tropics and subtropics	10–20	30–40	45–50
Desert plants			
Summer-germinating		20–30	
Winter germinating		10–20	ca. 30
Cacti		15–30	
Temperate-zone trees			
Conifers	4–10	15–25	35–40
Broad-leaved trees	below 10*	20–30	

* After cold-stratification.

The time and temperature needed for plant development can be quantified by calculating thermal units (TU) or degree days to cause an event as follows:

$$TU = \sum\left[\left(\frac{T_{max} + T_{min}}{2}\right) - T_{base}\right] \tag{9.1}$$

where T_{max} is daily maximum air temperature, T_{min} is daily minimum air temperature, and T_{base} is the base temperature. The base temperature is often set at about 10°C (50°F) for temperate plants (Fig. 9.5). The base temperature does vary among species. For example, wheat has a base temperature near 0°C, and peanut and cotton have base temperatures higher than 10°C. Also T_{max} values greater than 30°C (86°F) are entered as 30°C, since temperature response is curvilinear and asymptotic at about 30°C in most crop species.

Abbe (1905) used the term thermal constants for a given developmental stage of a plant receiving a specific quantity of TU, irrespective of time. This means that a specified quantity of TU is required for a developmental event, *e.g.*, the heading of wheat. Many modifications of the TU method have been made: nonetheless, the basic approach remains widely accepted.

Several factors or conditions can reduce the efficacy of the thermal unit method of expressing heat accumulation in relation to plant development.

1. It does not account for daylength which can greatly affect daylength sensitive crops such as soybean. Nuttonson (1953) included daylength in his heat accumulation system. For adapted genotypes and normal planting dates photoperiod may not present a problem and can be omitted in the TU system. The TU values obtained, however, may not be usable at other latitudes or planting dates.
2. Topography, aspect, and drainage can produce microclimates that do not reflect weather data from which thermal units are calculated.
3. Temperature extremes and drought can cause changes in developmental rate that are not accounted for in the thermal unit system. Also, the TU system uses air temperature, not plant temperature.
4. Crop pests and storm perturbation can alter the thermal unit system.
5. Irradiance in addition to temperature may alter the outcome of a developmental event. Some systems have combined irradiance and temperature and report units of development as photothermal units.

Plant Growth and Temperature

Photosynthesis

Like other growth processes photosynthesis is dependent on temperature. Photosynthesis increases as temperature increases up to an optimum temperature, depending on species and cultivar. Berry and Bjorkman (1980) observed in California that the optimum temperature for *Altriplex globriuscula*, a C_3 adapted to the coastal area, was about 25°C. On the other hand, the optimum

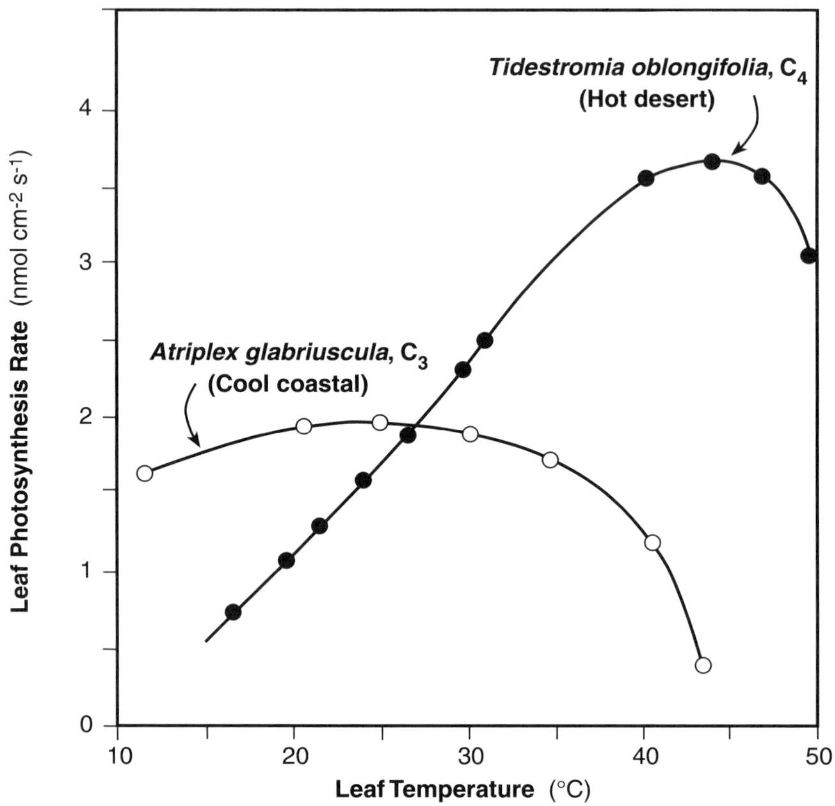

Figure 9.6 ***Leaf photosynthesis rates plotted against temperature for two plant species*** *(Berry and Bjorkman, 1980).*

temperature for *Tidestromia oblongifolia*, a C_4 adapted to inland deserts, was about 45°C (Fig. 9.6). Warm-season C_3 and C_4 species have optimum temperatures for photosynthesis of 25–35°C, and 30–40°C, respectively. The optimum temperature for cool-season C_3 species is about 20–30°C.

At stressfully high temperatures, about 5–10°C above the optimum, leaf photosynthesis may show time-dependent damage, being decreased progressively with continued exposure to high temperature. Chill injury on photosynthesis of chill-sensitive species can also result from low night temperatures, possibly because low night temperature interferes with the normal daily rhythms of biochemical processes. Species not only vary in their cardinal temperature for photosynthesis reactions, but they appear to have varied capacities to acclimate to cooler or warmer temperatures.

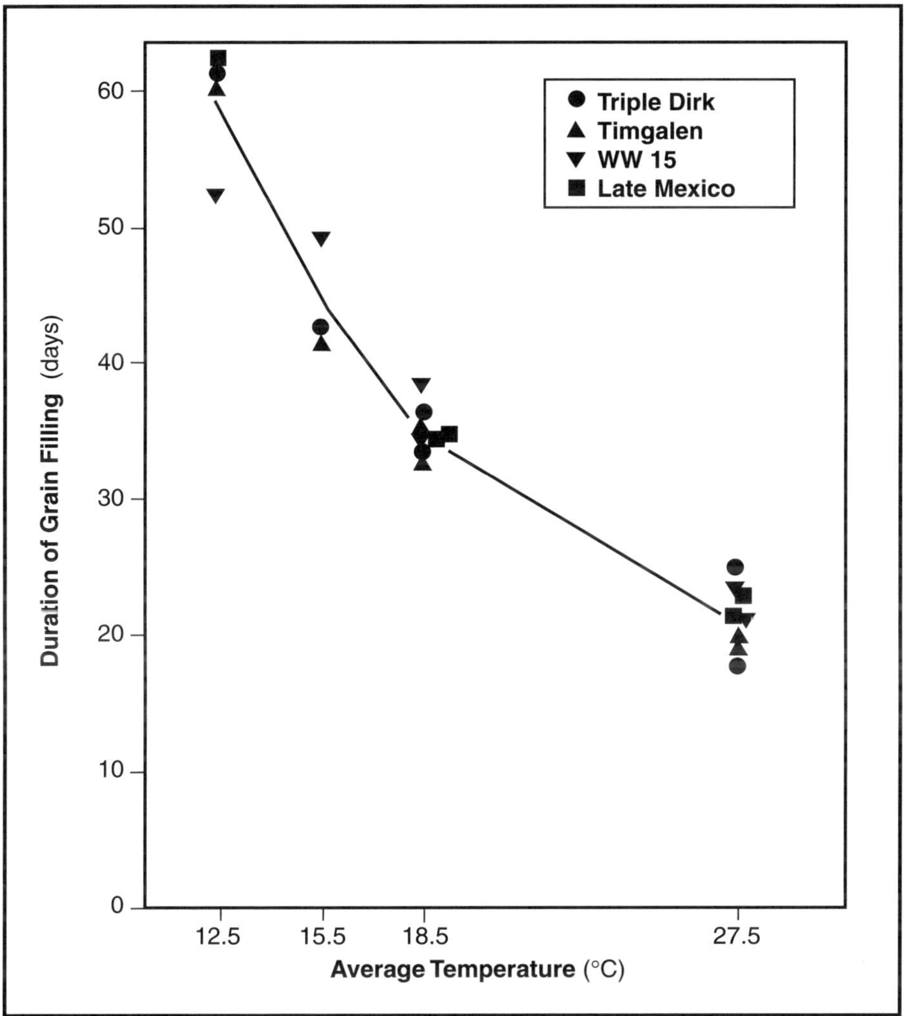

Figure 9.7 ***Duration of grain filling for four wheat varieties plotted against the temperatures at which the plants were grown*** *(redrawn from Sofield* et al.*, 1974).*

Duration of grain filling

Length or duration of seed or grain filling period is also important to yield potential and is inversely affected by increasing temperature. For example, the grain filling period of wheat is dramatically shortened as temperature increases (Fig. 9.7). Assuming 100% of the photosynthate goes to grain and no change in daily photosynthesis, wheat yield will decrease in direct proportion to shortening of the grain filling period as temperature increases.

The temperature-related effect on grain filling period is a major reason that wheat yield potential is much higher in northern Europe than in the Great

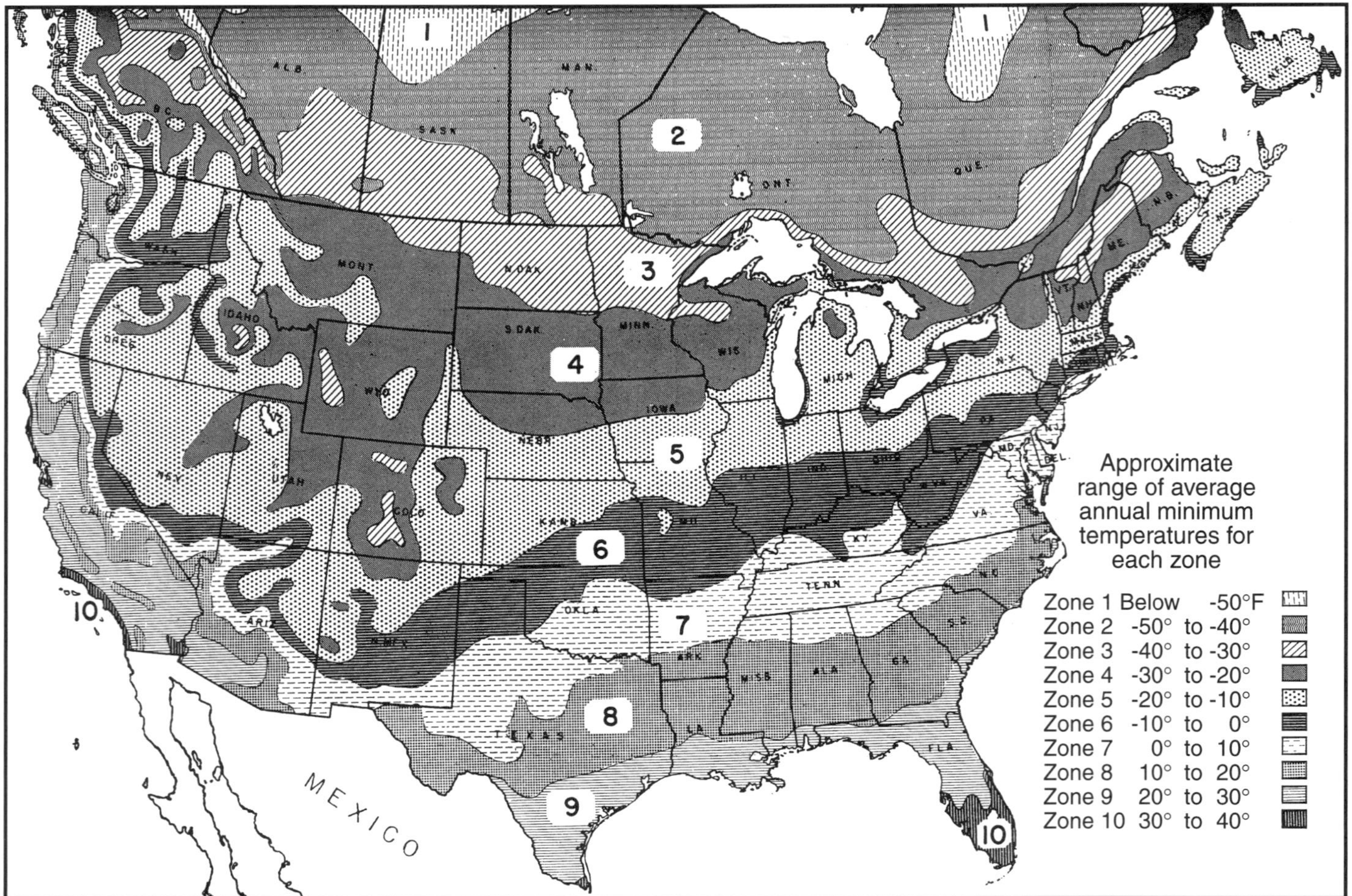

Figure 9.8 ***Zones of plant hardiness for the United States and Canada*** *(Wilsie, 1962).*

Plains of the USA. There have been attempts to lengthen the grain-filling period genetically. Yield improvement of maize hybrids in the USA during three eras (the 1930s, the 1950s and the 1970s) has, in part, been associated with an increase in duration of the grain-filling period. Length of the grain-filling period accounts for much of the yield variation among soybean cultivars.

Adverse temperature effects

Temperature outside the tolerance range of a species can severely damage the plant by protein denaturation and/or membrane damage. Freezing injury generally occurs inside the cells of the plant. The basis for the injury is physical rupturing of membranes by ice crystals, or desiccation of proteins by water extraction during ice-crystal formation. Marked changes in membrane chemistry, including reduction of size of proteins, may occur at sublethal temperatures.

Species vary widely in cold tolerance. Most temperate trees tolerate occurrences of severely low temperatures (< −40°C). On the other hand, subtropical trees (*e.g.*, citrus) can tolerate temperatures at 0 to −3°C for only a few hours, depending on level of cold-hardening. Except conifers, temperate perennials are deciduous. Some species are chill injured before freezing occurs. The US Department of Agriculture identified ten winter hardiness zones of the USA and Canada (Fig. 9.8). Tropical plants are adapted to Zone 10, whereas only extremely hardy species are adapted to Zone 1.

Plant parts and species differ in tolerance to low temperature. For example, a young fir tree is resistant to about −50°C, but actual tolerance temperature varies among parts. The cambium was much more resistant to cold (−80°C) than wood and leaves (−27 to −35°C). Plants have resistance and avoidance mechanisms to avoid heat and cold injury. Deciduous plants drop their leaves, while their young buds are covered with several layers of bracts or scales and have high osmotic concentrations. Buds and seeds of many temperate species have dormancy mechanisms and do not grow until certain temperature requirements are satisfied. Some perennials, e.g., alfalfa, die back completely to a crown submerged below the soil where temperature is more moderate and regrowth occurs from the dormant buds. Also, perennials have hardening mechanisms and other physiological traits to aid in cold survival.

Temperatures do not have to be below freezing (0°C) to damage tropical plants. Tropical plants may not germinate below 10°C. Chilling between 0 and 10°C was observed to cause membrane damage and leachates to move into the solution. Cacao seeds were severely damaged when chilled at 4°C for 10 minutes. Cotton seedlings were injured by chilling at 5°C. Effects of successive chilling of cotton seedlings at 10°C were cumulative in causing stunted growth. Cold soils also decrease seedling growth despite warm aerial temperature. Water uptake by roots may be inhibited at temperatures below 10–12°C.

Summary

Temperature influences the physiological processes of life, determines adaptation and distribution of species, and sets the practical limits for successful plant production. The temperature limits for growth are narrow (5–40°C) for crop species, but the range for survival is wide. Temperature for a given location is dependent upon season, latitude, altitude, and proximity to large bodies of water.

Heat is transferred as sensible (associated with change in temperature) and latent heat (no change in temperature but a change in state, *e.g.*, water to vapor). Conduction and convection (movement of parcels of air or liquid) are the primary processes of sensible heat transfer. Since the process of evaporation removes energy as latent heat, the evaporating surfaces and surrounding environment lose energy and are cooled. Evapotranspiration is the primary factor in the energy balance of areas covered by healthy vegetation.

Sandy soils are generally warmer than clay or organic soils primarily because of their lower moisture content and higher air content, resulting in lower diffusivity. Surface mulches reduce solar input and increase moisture retention, which results in dampening and lagging the sine-wave curve amplitude of temperature. Aspect and slope of the soil surface affect interception of direct solar radiation and resultant soil temperature.

Most species are adapted to warm day and cool night temperatures. The thermal units required for completion of a specific event (*e.g.*, flowering) vary with species. Some species require accumulation of cold units for completion of specific developmental events, *e.g.*, germination of some species, heading (vernalization), and breaking of bud dormancy (stratification). Many temperate perennial grasses require vernalization exposure to be coupled with short daylengths, but afterwards flowering of the induced plants requires long days. Photosynthesis and all growth processes are affected by temperature. Adverse low and high temperatures decrease the activity of various growth processes. Extreme temperatures obviously damage plants beyond recovery, although the temperature range of survival varies a great deal among plant species.

Further Reading

Abbe, C. (1905) *Report on Relations Between Climates and Crops*. USDA, Weather Bureau Bulletin 342.

Al-Nakshabandi, G. and Kohnke, H. (1965) Thermal conductivity and diffusivity of soils as related to moisture tension and other physical properties. *Agricultural Meteorology* 2, 271–279.

Berry, J. and Bjorkman, O. (1980) Photosynthetic response and adaptation to temperature in higher plants. *Annual Review of Plant Physiology* 31, 491–543.

Caborn, J.M. (1965) *Shelterbelts and Windbreaks*. Faber and Faber, London, pp. 48 and 76.

Chang, J. (1968) *Climate and Agriculture*. Aldine Publishing Company, Chicago.

Drozdov, S.N., Titov, A.F., Balagurova, N.I. and Kritenko, S.P. (1984) Effect of temperature on cold and heat resistance of growing plants. *Journal of Experimental Botany* 35, 1603–1608.

El-Sharkawy, M. and Hesketh, J. (1965) Photosynthesis among species in relation to characteristics of leaf anatomy and CO_2 diffusion resistances. *Crop Science* 5, 517–521.

Flood, R.G. and Halloran, G.M. (1986) Genetics and physiology of vernalization of wheat. *Advances in Agronomy* 39, 87–125.

Gardner, F.P. and Barnett, R.D. (1990) Vernalization of triticale and wheat and triticale cultivars. *Crop Science* 30, 166–169.

Gardner, F.P., Pearce, B.R. and Mitchell, R.L. (1985) *Physiology of Crop Plants*. Iowa State University Press, Ames, Iowa.

Geiger, R. (1965) *The Climate Near the Ground*. Harvard University Press, Cambridge, Massachussetts.

Jadel, P.E., Evans, L.E. and Scarth, R. (1986) Vernalization responses of a selected group of spring wheat (*Triticum aestivum* L.) cultivars. *Canadian Journal of Plant Science* 65, 33–39.

Khatibu, A.I., Lal, R. and Jana, R.K. (1984) Effects of tillage methods and mulching on erosion and physical properties of a sandy clay loam in an equatorial warm humid region. *Field Crops Research* 8, 239–254.

Kiniry, J.A. and Bonhomme, R. (1991) Predicting maize phenology. In: Hodges, T. (ed.), *Predicting Crop Phenology*. CRC Press, Boca Raton, Florida.

Larcher, W. (1980) *Physiological Plant Ecology*, 2nd edn. Springer-Verlag, New York.

Lee, R. (1978) *Forest Microclimatology*. Columbia University Press, New York.

Lemon, E., Stewart, D.W. and Shawcroff, R.W. (1971) The sun's work in a cornfield. *Science* 174, 371–378.

Levitt, J. (1978) An overview of freezing injury and survival, and its relationships to other stresses. In: Li, P.H. and Sakai, A. (eds) *Plant Cold Hardiness and Freezing Stress, Mechanisms and Crop Applications*. Academic Press, New York.

McCalla, R.M. and Dudley, F.L. (1946) Effect of crop residues on soil temperature. *Journal of the American Society of Agronomy* 37, 75–89.

Miller, A. and Thompson, J.C. (1979) *Elements of Meteorology*. Charles E. Merrill Publishing, Columbus, Ohio.

Nuttonson, M.Y. (1953) *Phenology and thermal environment as a means for physiological classification of wheat varieties and for predicting maturity dates of wheat*. American Institute of Crop Ecology, Washington, DC.

Rosenberg, N.J., Blad, B.L. and Verma, S.B. (1983) *Microclimate: the Biological Environment*. John Wiley and Sons, New York.

Shreve, F. (1924) Soil temperature as influenced by attitude and slope exposure. *Ecology* 5(2), 128–136.

Sofield, I., Evans, L.T. and Wardlaw, I.F. (1974) The effects of temperature and light on grain filling in wheat. In: Bieleski, R.L., Ferguson, A.R. and Cresswell, M.M. (eds), *Mechanisms of Regulation of Plant Growth*. Bulletin 12, Royal Society of New Zealand, Wellington.

Vemura, M. and Yoshida, S. (1986) Studies on freezing injury in plant cells. *Plant Physiology* 80, 187–195.

Wilsie, C.P. (1962) *Crop Adaptation and Distribution*. W.H. Freeman, San Francisco.

Wolk, W.D. and Herner, R.C. (1982) Chilling injury of germinating seeds and seedlings. *Horticultural Science* 17, 169–173.

Weather and Climate 10

L.H. ALLEN AND F.P. GARDNER

All terrestrial organisms live in a thin gaseous envelope that surrounds the Earth, the atmosphere. The collective characteristics of the atmosphere over many years define the climate. The climate determines the kinds of organisms that can inhabit a place. It sets the limits of adaptation and production of plants, and therefore sets constraints on plant production ecosystems. The climate determines the distribution and migration patterns of many animals. Only humans can drastically modify their 'microclimate' to suit their needs, particularly in industrialized countries with advanced technology.

A topic of common discussion, the weather is the state of the atmosphere at a specific time and place. Meteorology is the study of weather and the dynamics of specific atmospheric elements. The accumulated weather events over many years constitute the climate. Climatology is the study of the climate and its effects on organisms. The basic meteorological elements that form the climate include solar radiation, temperature, precipitation, winds, and humidity. These factors are influenced by latitude, altitude, and proximity to large bodies of water.

The climate can be discussed in terms of its principal determining factors: (i) solar radiation; (ii) albedo; (iii) atmospheric and oceanic circulation; (vi) mechanisms of precipitation; (v) spatial distribution of land and water; and (vi) physiographic land factors, especially altitude.

Solar Radiation

Solar radiation, discussed in Chapter 8, is the driving force of weather. As seen previously, its distribution over an area varies widely according to latitude and season. Global annual means of solar radiation vary about three-fold, consequently, large differences in climate prevail.

Albedo

Albedo refers to the fraction of the solar radiation reflected from a surface. About a third of the incident solar radiation is reflected by clouds compared with about 5% reflected by water surfaces and 5–25% by land and vegetation surfaces. Reflection from clouds varies depending on cloud type and on thickness of the cloud cover. A cloud thickness of a few meters reflects about 25% of the solar radiation, whereas a 1000-m thick cloud reflects about 90% of the solar radiation. Albedo is also influenced by: (i) surface characteristics (fresh snow albedo is five- to ten-fold greater than that of green crops or forest); and (ii) season (albedo is high in winter due to the low solar angle).

Some meteorologists attribute desertification associated with the recent long drought in the African Sahel, to increased albedo. Their reasoning is that serious losses of vegetation, because of overgrazing and drought, exposed light-colored desert soils and rocks. Consequently, the albedo of the region was increased and there was less heating of soil and air. The decreased warming allowed for the development of cool dry air in the atmosphere. The density of the cool air is greater resulting in desert, high-pressure cells. These atmospheric cells expand, causing precipitation to shift elsewhere and resulting in desert expansion.

Atmospheric Circulation

Vertical and horizontal movement of the atmosphere due to winds or eddies distributes energy over the Earth and activates other weather components. Wind is a response to three factors: (i) pressure gradients resulting from differential heating and cooling; (ii) gravity causing denser air parcels to sink and displace lighter air parcels (buoyancy); and (iii) centrifugal force resulting from the Earth's rotation, the Coriolis effect.

Coriolis effect

Winds do not directly cross latitudes at right angles. Winds move obliquely across latitudes because of the centrifugal force caused by the spin of the Earth, i.e., the Coriolis effect (Fig. 10.1). In the northern hemisphere, the wind deflection is clockwise, but in the southern hemisphere it is counter-clockwise.

Cyclones and anticyclones

Winds, developing in response to low pressure cells, flow inward toward the center of the cell and are termed *cyclones*. Winds from high temperature (and pressure) cells, flow outward and are termed *anticyclones*. Trade and westerly winds (Fig. 10.1) are anticyclonic since they flow from the large high-pressure

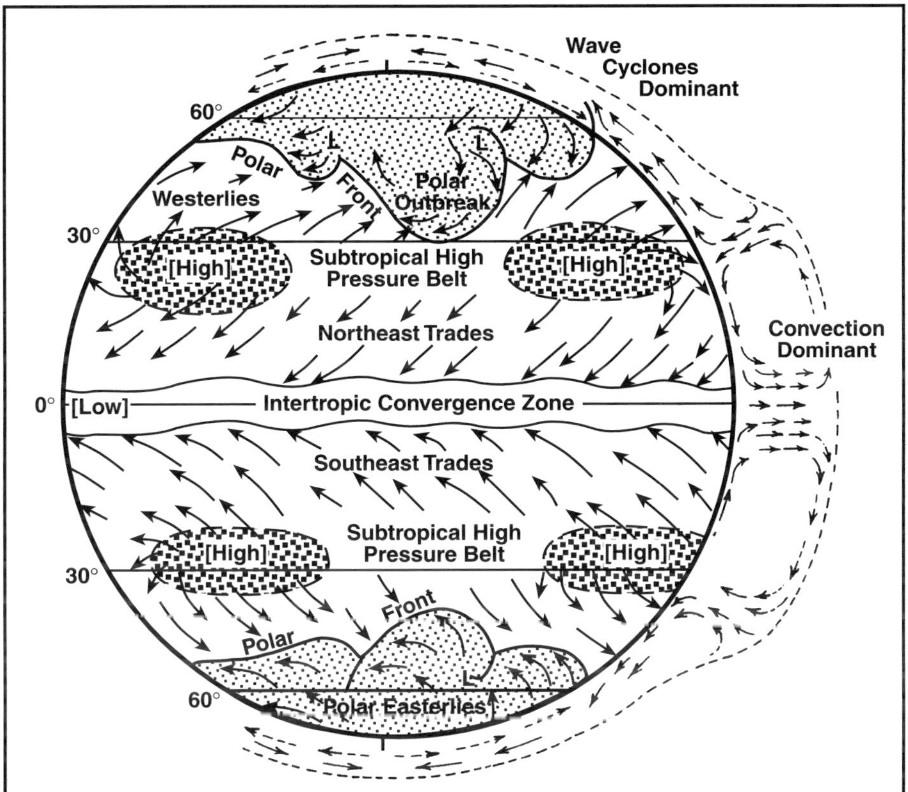

Figure 10.1 ***Schematic of global wind patterns*** *(redrawn from Strahler and Strahler, 1978).*

cells of the subtropics. These winds are responses to both pressure and the Coriolis effect. In the northern hemisphere, the westerlies are clockwise and therefore tend to circumvent the eastern periphery of the high-pressure cells. The result is the cool, dry air in summer of the popular Mediterranean climate, as is experienced on the west coast of the USA and in southern Europe. The reverse is true for the trade winds and the result is warm, humid weather as experienced in the southeast USA, Southeast Asia, and the Philippines. Westerlies also meet the polar easterlies to produce storms and wet weather in mid-latitudes, in effect a convergence zone like that at the equator. The deeper continental interiors of these permanent high pressure cells are low in rainfall.

Global surface winds

Standard atmospheric pressure at sea level is 1013 millibar (mb), but atmospheric pressure may vary from a low of about 980 mb to a high of about 1040 mb. If there were no differences in atmospheric pressure, the condition would be isobaric as may be the case for a given time and area. However, large and permanent high pressure belts are generally located around the globe in both northern and southern hemispheres at about 20–30°. Their locations shift somewhat with the movement of the sun between summer and winter, i.e., they follow the high sun. The zone in the northern hemisphere moves northward from the equator until the summer solstice. Similarly, in the southern hemisphere the zone moves southward from the equator in the summer of the southern hemisphere. As will be seen later, this factor is the determinant of the wet and dry seasons lying between 10 and 20° N and S (the savannas). Low pressure zones also occur at about 50–60° N and S (the polar fronts). These also respond to the track of the sun, like those at 20–30° N and S.

The northern polar front can reach southward to 30° N or farther (polar outbreak). This results in a much colder and more variable climate in temperate latitudes during winter in the northern hemisphere. In the southern hemisphere the polar front does not extend as far from the pole; the Earth's surface is covered mainly by oceans in the southern hemisphere that do not freeze, whereas the northern hemisphere has much more land mass surrounding the Arctic Circle. The land mass of Antarctica blocks poleward transport of heat by ocean currents, whereas the Arctic receives heat from ocean currents (the Gulf Stream) continually. These factors make the south circumpolar vortex more stable than the north circumpolar flows. This stable circumpolar vortex over Antarctica is one of the meteorological factors that contributes to the development of the 'ozone hole' at the end of the southern hemisphere winter.

The polar outbreaks in the northern hemisphere restrict the cultivation of tropical and subtropical crops to low latitudes in the USA compared with South America. Citrus can be grown successfully at latitudes no greater than about 28° N in Florida in contrast with production at 35° S in South America. The polar air outbreaks during the winters spanning 1977–1985 froze out most of the citrus industry in the central peninsula of Florida (28–29° N). This led to the development of new citrus groves farther south in that state.

Without fronts, winds move from high to low pressures zones. Examples of some important winds are those that originate from the subtropical highs ('horse latitudes', 20 to 30° N and S). Winds move from these highs toward the subpolar front (60°) as westerly winds and toward the equator as easterly trade winds. In the early days, ships leaving North America could be rapidly pushed home to Europe by sailing north and catching these westerlies. Westerlies converge with polar winds at the polar front. The prevailing surface trade winds converge at the intertropic convergence zone, known also as the equatorial trough and doldrums. As a result there is a band of updrafts near the equator (Fig. 10.1). The systems of trade winds and doldrums have tremendous effects on tropical weather. They form the climatic resource (particularly rainfall) for the production of cereal crops, especially rice, in this intertropic convergence zone. The average annual rainfall in the intertropic convergence zone of the Pacific Ocean can be as high as 5000 mm.

Monsoon winds

Differential heating/cooling of water and land masses creates large thermal cells in the atmosphere, *i.e.*, air circulation due to a temperature differential (Fig. 10.2). These thermally driven atmospheric circulation cells, or monsoons, exist on all continents but the best known are those in Southeast Asia and India. Summer winds laden with moisture from the warm Indian Ocean and the southwest Pacific move northwestward over the land masses of India, Bangladesh, Southeast Asia, and China. As the air carried by these winds rises and cools over land, the moisture condenses and falls as monsoonal rains (Fig. 10.2a). The monsoons provide the rainfall for crop

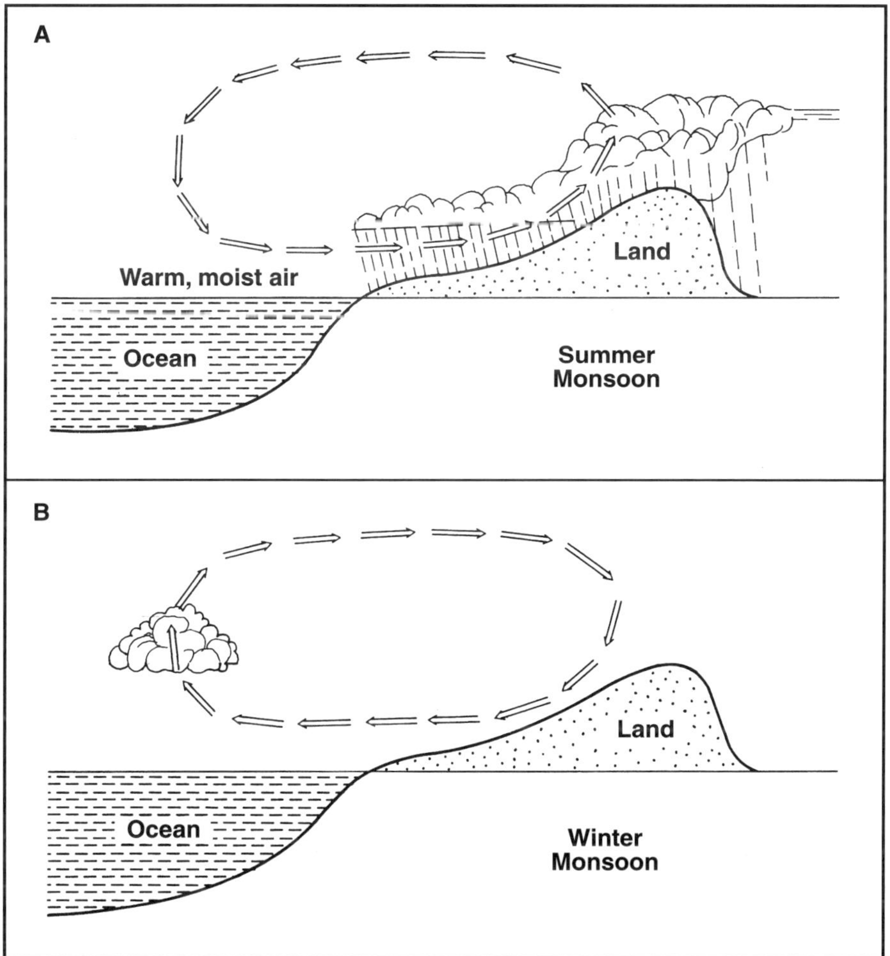

Figure 10.2 ***Schematic of weather patterns in (A) summer and (B) winter for monsoonal climates*** *(redrawn from Miller and Thompson, 1979).*

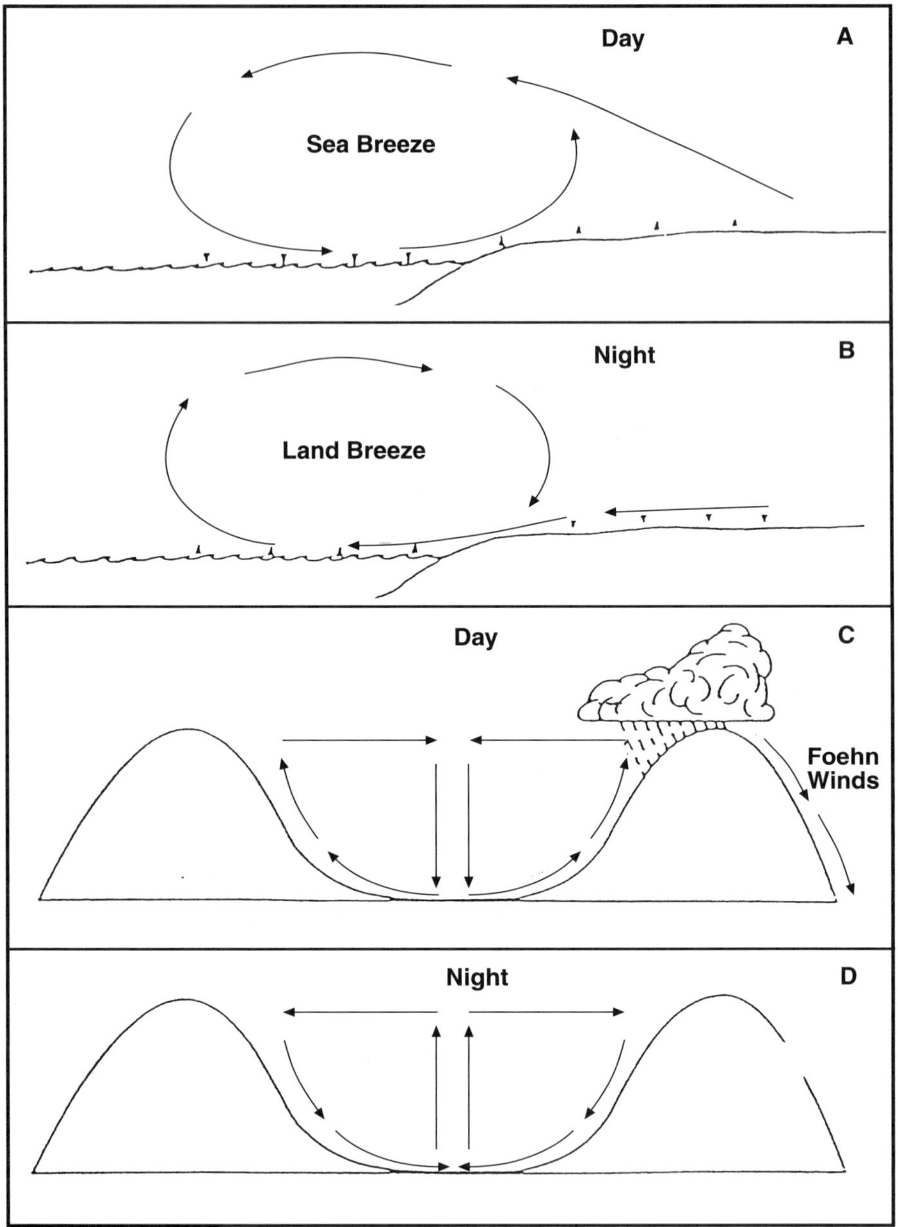

Figure 10.3 ***Schematic of local wind patterns for (A,B) coastal areas, and for (C,D) mountain and valley areas.***

production and human livelihood for billions of people who live in the vast area of land south and east of the Himalayan mountains. The highest monthly rainfall recorded on Earth is 9300 mm at Cherrapunja, India, in July 1861. This location also recorded the highest annual rainfall, 26,000 mm.

In winter, the monsoons are reversed because the oceans are warmer than the land and, therefore, rainfall is slight to nil (Fig. 10.2b). Although less pronounced, a similar pattern of monsoons exists over the Gulf of Mexico and eastern USA. The summers are humid and moist due to winds from the Gulf. Thrall, Texas received 955 mm of rain on 9–10 September 1921. Winters tend to be dry in monsoon areas. When these monsoons fail, as they sometimes do, drought, food shortages, and even dire famines occur in populations dependent on the monsoon rains.

Local winds

Besides large scale pressure systems and circulation, as described above, local surface winds arise from thermal differentials in small areas of the immediate surrounding terrain. Sea breezes result from rapid heating of land in the day causing the warmed surface air to rise. Winds carry cool air from above the water (seas or lakes) to displace the lighter rising air over the land (Fig. 10.3). At night the opposite happens as the land cools faster than the water and thus surface winds move from land toward water. Sea breezes produce desirable cooling effects (for the human populace) and help in dissipating polluted air over urban areas near the sea.

Mountain winds develop when valleys are warmed during the day, air rises, and moves up the mountain slopes to the summit (Fig. 10.3). At night, the valleys are cooled by loss of thermal radiation and the air movement is reversed, *i.e.*, down the slope. Furthermore, cool air is denser or heavier than warm air and, therefore, it sinks from the higher to lower levels at night due to gravity. Another common mountain breeze is the Foehn, or Chinook wind. Rising air masses from the valley are often cooled to their dewpoint, the temperature at which atmospheric water vapor condenses to liquid, as they move up the mountain slope and produce precipitation at or near the summit (Fig. 10.3). This change in state from gas to liquid releases energy so the air is warmed. The expended and consequent dry air is pushed down the (usually) opposite slope where it is further warmed. The result is warm, dry winds that move briskly down mountain slopes. An example of these winds is those of the western mountain range of California, which are called the Santa Ana winds in the Los Angeles area.

High altitude winds

Winds several kilometers above the Earth, including polar fronts, tropical winds, and jet streams have great effects on weather. Polar jet streams can bring severe cold to a region, and tropical jet streams push moist humid air into a region.

Fronts, cyclones, and precipitation

Large circular movements play an important role in determining weather. In the middle latitudes, cyclonic depressions occur that emanate from the polar fronts (Fig. 10.1). Contact of dense polar air with warm, less-dense air may establish a shear and cyclical motion. Large warm- and cold-air masses in these waves form fronts and are responsible for much of the precipitation in mid-latitudes. As the warm fronts meet the cold fronts, precipitation occurs.

Tropical cyclones, as their name implies, form over the tropics, usually between 5° and 20° latitude. Fully-developed tropical storms are termed hurricanes in North America, typhoons in Asia, and various other terms in other places. Low pressure cells form beneath high clouds. As the clouds become lower, wind speed increases and, near the center, speed becomes severe causing heavy to violent rainfall.

Precipitation occurs when warm, moist air is cooled to its dewpoint. The dewpoint temperature is where water vapor in the air condenses into the liquid phase. Nearly all precipitation that falls results from cooling of rising air. Three mechanisms resulting from this process are described.

1. Orographic precipitation occurs when a warm moist air mass is lifted by mountains resulting in cooling of the air. Fog and rain zones near the summit of mountains are common in the tropics (Fig. 10.4A). After the release of water, air is warmed and may descend the opposite slope of the mountain range as Foehn winds (Fig. 10.3). The heavy snows of the Sierra Nevada mountains of California are orographic precipitation.

2. Frontal precipitation results when warm air masses are lifted over the top of cooler and denser air masses, resulting in condensation and precipitation (Fig. 10.4B). Precipitation in middle latitudes is generally from frontal storms, especially winter precipitation.

3. Convectional precipitation occurs when warm air rises, forms cumulonimbus clouds, and condensation occurs (Fig. 10.4C). Thunderstorms are, in effect, the result of such local thermal cells. As warm air rises, condensation occurs and cumulus clouds form as the air is cooled to the dewpoint temperature. As air is convected higher, water and/or ice coalesces and falls as precipitation. The descending wind from these cells is cool and very strong in severe storms. These downbursts can be particularly hazardous for aviation.

Climate

Climates are produced by the mean composite variations of the weather elements as discussed above. Climate and zones of precipitation are linked to the prevailing global circulation patterns as illustrated in Fig. 10.1. The converging surface winds and concomitant towering updrafts produce the intertropic convergence zone (doldrums) near the equator (0° ± 10°). This zone is characterized by a band of thunderstorms that circumscribe the Earth. Astronauts have

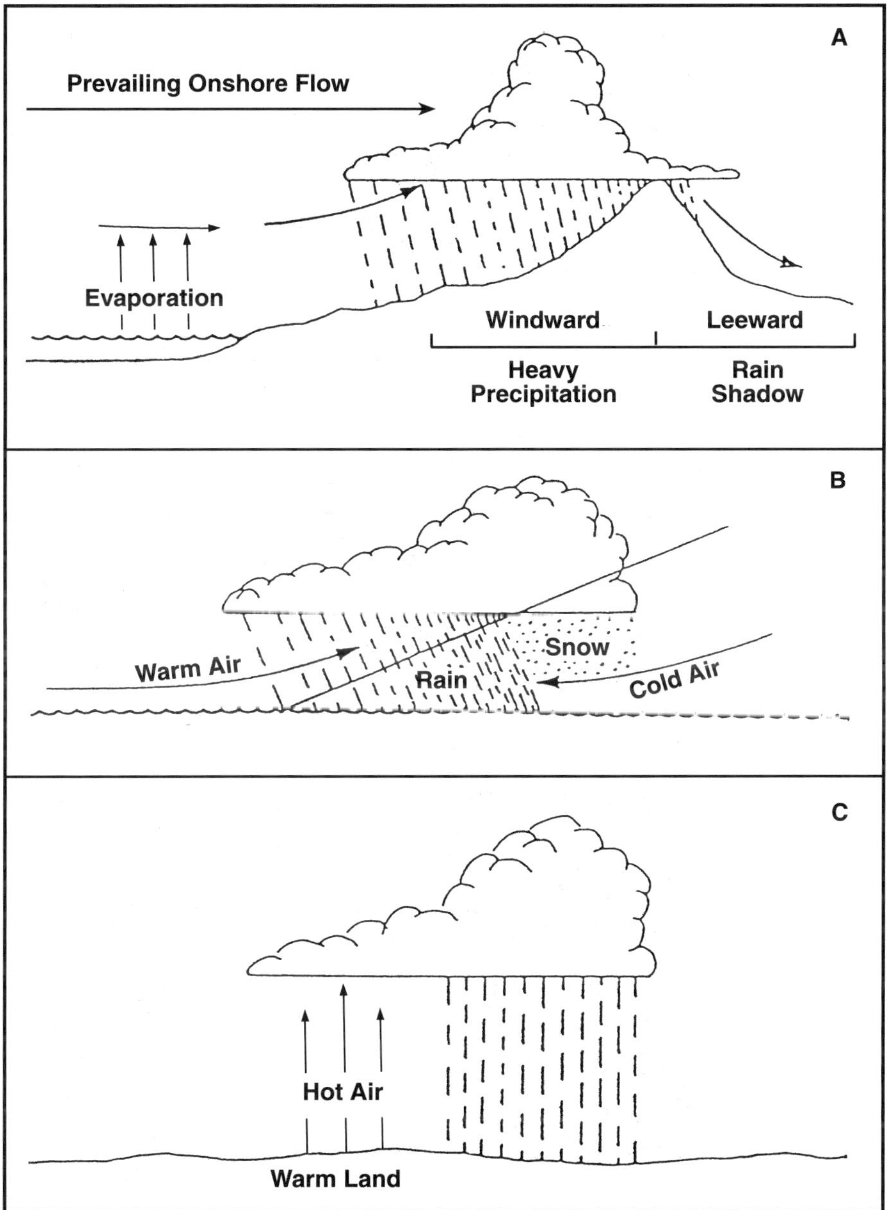

Figure 10.4 ***Schematic of precipitation resulting from (A) elevation changes of moist air, (B) frontal system of warm and cold air, and (C) convection of warm, moist air.***

observed almost continuous lightning from thunderstorms in this zone. The intertropic convergence zone shifts northward or southward from the equator during the northern hemispheric summer or winter, respectively. This seasonal shift in the latitude of the intertropic convergence zone produces seasonal

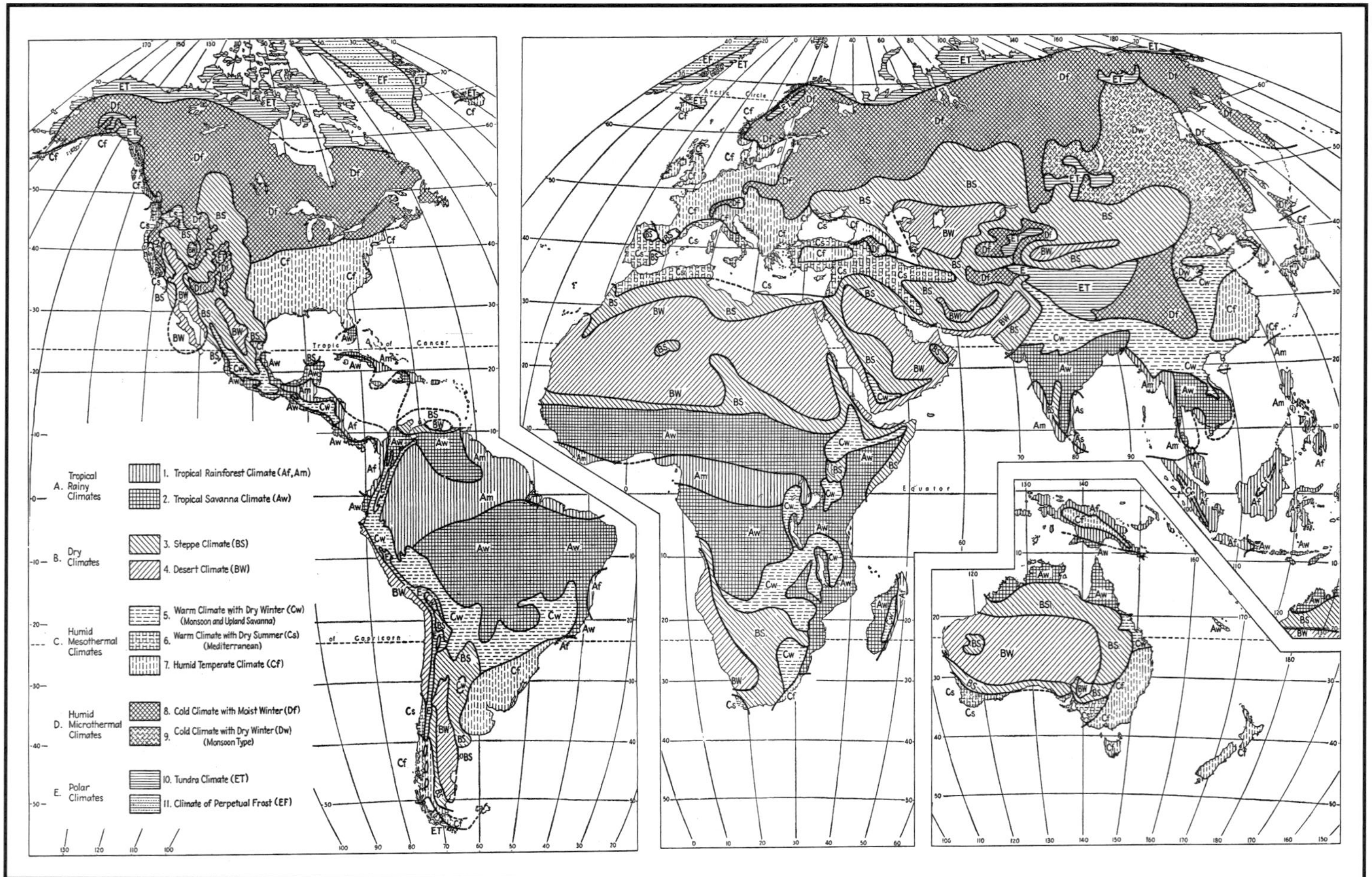

Figure 10.5 ***Global climate classifications as proposed by Thornthwaite*** (Janick et al., 1974).

variations in precipitation. That is, this zone is wet in summer and dry in winter of each hemisphere, which consequently causes savannah-type climates. Near the equator (0° ± 5°) the intertropic convergence zone overlaps both northern and southern hemispheres and as a result, rain falls in all seasons. However, the amount of rain near the equator also varies from month to month. A similar convergence effect occurs at 60° N and S where the westerlies converge with polar winds resulting in precipitation in all seasons (Fig. 10.4b).

Climate classification

Many classification schemes of climate have been published. The oldest and probably the best known is the one by Koppen, an Austrian geographer, for whom it is named. The system is based on changes of vegetation. Its weaknesses are that it is not based on specific meteorological data, and transitional areas are not well defined. For example, by the Koppen system Florida is said to have a climate like Maryland and the mountains of North Carolina. Koppen's tropical desert and Mediterranean climates were better defined. The mild winter and cool dry summer of the Mediterranean climate is considered ideal for humans.

Thornthwaite (1931) recognized that the absolute values of precipitation and temperature are not as important as the effective values of these parameters. He developed a climate classification system based on effectiveness of primary weather elements. The Thornthwaite classification (Fig. 10.5) is easy to use and includes meteorological parameters for better definition. It could be improved by including wind, humidity, and other aerodynamic data.

Climate Change

Many people are not only concerned with the current climate, but also with the climate of the future. The phenomenon of the 'greenhouse effect' (see Chapter 11) has greatly heightened this concern. According to some projections, the Earth could warm by as much as 3–5°C due to a continuing elevation of CO_2 concentration in the atmosphere. Records currently show a warming trend, especially over land masses in the southern hemisphere. Warming of the Earth by 3–5°C would affect climate, including amount and distribution of rainfall, ice caps, and sea levels. Current crop production centers might shift to different latitudes, but the new latitudes may not present the best topography, soils, and other environmental resources for crop production. The science of paleoclimatology (prehistoric) has employed a number of techniques to develop a fairly complete climatic record for the past centuries up to as long ago as 10,000 years. These techniques have included carbon dating and studying tree rings, ice layers, and layered lake sediments. This record has been extended to the past 160,000 years based on ice cores obtained from Antarctica and Greenland, and coral accumulation from various seas. Going back in time, the major events in this record are summarized as follows (Fig. 10.6).

1. A cold period existed from 1880 to about 1900. This was followed by warming that peaked in about 1940 (Fig. 10.6A).
2. The period from AD 1430–1850 (Fig. 10.6B) was identified as the 'little Ice Age' and was preceded by a warm period of several hundred years. During the little Ice Age, many glaciers of Scandinavia, Alaska, and the Alps advanced near to their position during the major Ice Age. Expansion of the Arctic ice pack caused the Norse colony of southwest Greenland to perish. In Iceland, grain crops that had been grown for centuries could no longer survive.
3. During the last 15,000 years, the period 7000–5000 years ago was interglacial and warmer than it is today. The Younger Dryas cold interval, which ended about 10,000 years ago (Fig. 10.6C), was followed by the development of ancient civilizations.
4. A major Ice Age persisted to varying degrees from 15,000 to about 115,000 years ago, with the coldest periods being about 15,000–40,000 years before present (Fig. 10.6D). Ice sheets extended and receded over the northern hemisphere. Northern forests from previous warm periods were destroyed and at times sea level was about 100 m lower than it is today. Migrations of animals and plant communities took place.

Summary

The climate is a critical determinant in the adaptation and distribution of living organisms. This is especially true of plant production ecosystems that must not only survive but produce economically. The aggregation of the weather elements over many years, such as solar radiation, temperature, precipitation, winds, and humidity, forms the climate of a specified location. Climatology and meteorology are studies of these factors. Major climate determinants include solar radiation, albedo, Earth surface characteristics, atmospheric and oceanic circulation, and precipitation mechanisms. Precipitation as rain or snow results when warm, moist air parcels are forced to rise and cool to the dewpoint temperature. Climates have been classified by several climatologists. The oldest and best known of these classification systems is the Koppen, which is based on the assumption that vegetation type indicates climate type. The Thornthwaite system is based on temperature and precipitation effectiveness, as computed by his models, and their relationship to characteristic vegetation associated with the temperature and humidity types. The climate we now have has not always existed. Paleoclimatological studies have shown clear variations in the Earth's climate in past ages.

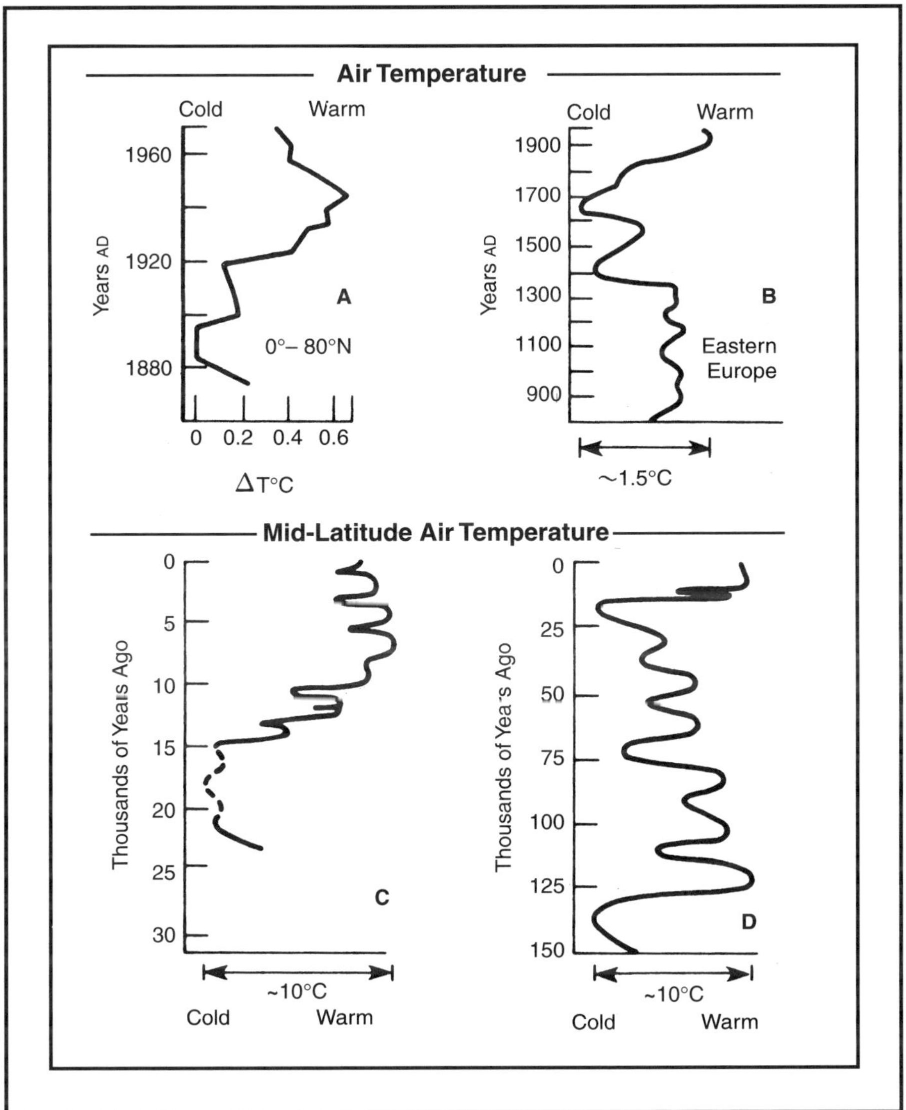

Figure 10.6 ***Historical air temperatures in the northern hemisphere for increasingly longer time spans*** *(Pollack, 1982).*

Further Reading

Akin, W.E. (1991) *Global Patterns: Climate, Vegetation, and Soils*. University of Oklahoma Press, Norman, Oklahoma.

Anon. (1988) Climatic variability. *Southern Climate Review* 1, 18–26.

Cox, G.W. and Atkins, M.D. (1979) *Agricultural Ecology, an Analysis of World Food Production*. W.H. Freeman, San Francisco.

Gates, D.M. (1972) *Man and his Environment: Climate*. Harper and Row, New York.

Goodess, C.M., Palutikos, J.P. and Davies, J.P. (1992) *The Nature and Causes of Climate Change: Assessing the Long Term Future*. CRC Press, Boca Raton, Florida.

Janick, J. Schery, R.W., Woods, F.W. and Ruttan, V.W. (1974) *Plant Science – an Introduction to World Crops*, 2nd edn. W.H. Freeman, San Francisco.

Lutgens, F.K. and Tarbuck, E.J. (1995) *The Atmosphere: an Introduction to Meterorology*. Prentice Hall, Englewood Cliffs, New Jersey.

Manabe, S. and Broccoli, A.J. (1984) Ice-age climate and continental ice sheets: some experiments with a general circulation model. *Annals of Glaciology* 5, 100–105.

Miller, A. and Thompson, J.C. (1979) *Elements of Meteorology*. Charles E. Merrill, Columbus, Ohio.

Muchow, R.C. and Bellamy, J.A. (eds) (1991) *Climatic Risk in Crop Production: Models and Management for the Semiarid Tropics and Subtropics*. CAB International, Wallingford, UK.

Pollack, J.B. (1982) Solar, astronomical and atmospheric effects on climate. In: *Climate in Earth History*. National Academy Press, Washington, DC.

Postel, S. (1989) Land's end. In: *World Watch* 3, 12–20.

Strahler, A.N. and Strahler, A.H. (1978) *Modern Physical Geography*. John Wiley and Sons, New York.

Thornthwaite, C.W. (1931) The climates of North America according to a new classification. *Geographical Review* 21, 633–655.

Carbon Dioxide and Other Atmospheric Gases

11

L.H. ALLEN

Modern human activities cause the release of large quantities of gases into the atmosphere, potentially affecting the quality of the environment. Plant and animal health may be affected directly by ozone and oxidants produced in the atmosphere. These gases result from a combination of hydrocarbons from various sources and oxides of nitrogen from automobile exhausts. In addition, emission of industrial gases can affect climate and can change the chemistry (pH and ionic content) of precipitation. Both acidic rain and dry deposition of these emissions can enter bodies of water and impact aquatic biota adversely, especially in poorly buffered bodies of water.

On the other hand, gases of carbon (CO_2), nitrogen (NO_2 and NH_3), and sulfur (SO_2) contain essential nutrients that may benefit plant growth. The primary focus of this chapter is on carbon dioxide (CO_2), because of the quantity emitted, its direct role in photosynthesis and plant growth, and its role as a greenhouse-effect gas. CO_2 is of major ecological significance because it affects plants both directly and indirectly through its role in governing the Earth's surface temperature.

About 0.036%, or 360 μmol mol^{-1}, of dry atmospheric air is CO_2 (on a molecular basis) while 78% is N_2 and 21% is O_2. Although CO_2 concentration is low, it is very important to plant and animal environments because: (i) CO_2 is an essential plant nutrient for photosynthesis; (ii) CO_2 is a bioregulator of stomatal conductance and leaf gas exchange; and (iii) atmospheric CO_2 helps regulate the radiation energy balance and temperature of the Earth.

Carbon dioxide and certain other trace gases (CH_4, N_2O, O_3, and chlorofluorocarbons) absorb (and emit) thermal infrared radiation in various bandwidths across the 3–100 μm wavelengths. Thus, as energy is emitted from the Earth's surface, some of it can be absorbed by CO_2 and other thermal radiation absorbing gases. After this energy has been absorbed, it has an equal probability of being reradiated downward and upward. This process of absorption and reradiation can recur repeatedly so that a substantial portion of the thermal radiation returns to the Earth's surface. Consequently, the Earth's surface is prevented from being cooled to the extent that would occur if there were no

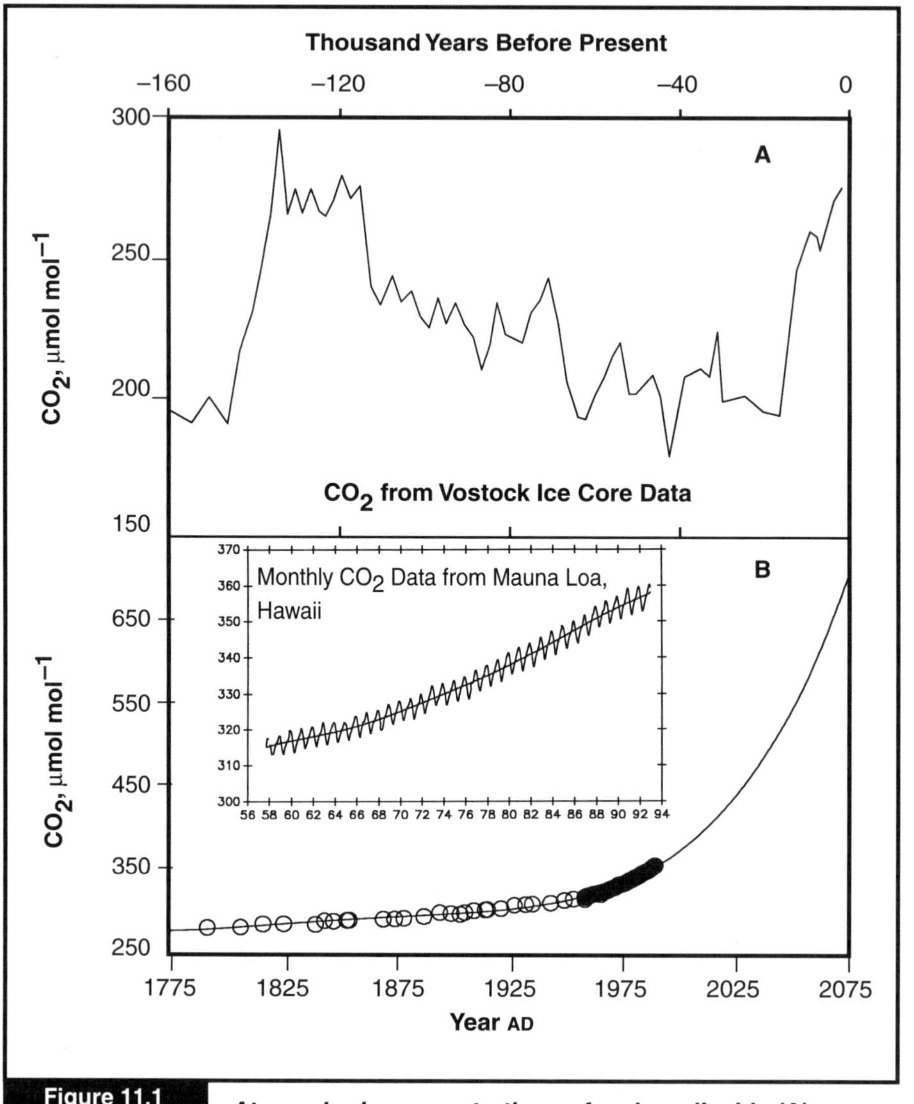

Figure 11.1 ***Atmospheric concentrations of carbon dioxide (A) past 160,000 years, and (B) over past 200 years with a projection for the future. The insert in panel (B) shows carbon dioxide within the annual cycle for recent years.***

atmosphere with greenhouse gases. This process has been called the 'greenhouse effect' because the incoming solar radiation is transmitted through the transparent atmosphere to the Earth's surface and the outgoing thermal radiation is absorbed in the atmosphere by certain gases. The Earth system exists in thermal equilibrium over the long term. That is, incoming solar radiation energy at the outer limits of the atmosphere must be balanced by the same amount of total outgoing radiation energy at the top of the atmosphere.

The concentrations of the greenhouse-effect gases in the atmosphere provide the Earth with an average climate at the surface that is about 15°C (59°F). This, of course, is warmer than the effective average surface temperature would be without the blanket of atmospheric gases. Without the greenhouse gases in the atmosphere, the average temperature at the Earth's surface has been calculated to be about −18°C (0°F). Therefore, the existing levels of greenhouse-effect gases are essential for life on Earth as we know it. However, if these gas concentrations rise (and they are increasing), this increase would be expected to cause additional global warming at the Earth's surface.

The atmospheric concentration of CO_2 is increasing at about 1.5 μmol mol^{-1} per year as shown in Fig. 11.1 and is projected to double in concentration by about 2075. Such a CO_2 increase has a potentially serious consequence in global warming. The CO_2 concentration rose from 265 to 280 μmol mol^{-1} in the Holocene, pre-industrial revolution times. Since the beginning of the industrial revolution the CO_2 concentration has risen to the current level of about 360 μmol mol^{-1} (measured at Mauna Loa Observatory, Hawaii). Monitoring of CO_2 in Australia from 1972 to 1981 showed a trend similar to that at Mauna Loa.

Global CO_2 Cycle

Nearly all energy that sustains life of terrestrial, marine, and soil organisms comes directly or indirectly as a result of respiration of photosynthetic products: ($(CH_2O)_n + O_2 \rightarrow CO_2 + H_2O$ + energy). Therefore, Earth biota, especially roots, heterotrophic plants, microbes, and fauna are important sources of CO_2 in the global carbon cycle. Forests are especially important because they are 90% of the living plant mass although they cover only 30% of the Earth's surface. Consequently, forests are a major component in the global carbon balance. The contribution of CO_2 to the atmosphere from forests of developing countries (mostly tropical) far exceeds that in developed countries. All CO_2 from respiring heterotrophic plants and fauna is released into the atmosphere. However, it is estimated that 30–40% of the amount of CO_2 released in respiration of autotrophic plants is recycled in photosynthesis, excluding dramatic perturbations such as burning and logging operations. The estimate of net carbon released by respiration of disturbed terrestrial ecosystems ranges from $0.6–2.6 \times 10^9$ metric tonnes of carbon per year.

Effect of soil on the carbon balance

Much of the rise in CO_2 concentration from about 275 (pre-industrial levels) to about 300 μmol mol^{-1} (in 1940) can be attributed to widespread development of agriculture. These activities were primarily clearing of forests, ploughing of the prairies, and cultivation of soils, especially those high in organic matter. Cultivation stimulated biological activity in the soil, which caused rapid loss of organic matter in labile forms (Fig. 11.2). The 1–2 × 10^6 metric tonnes of carbon per year estimated to be released from forests and soils is about 2–4% of net terrestrial primary production. While deforestation seems to continue unabated globally (about 7 × 10^6 hectares per year), the absolute emission from forest and soils might be declining while that from fossil fuels continues to escalate (Fig. 11.2).

Cement manufacture

Cement is manufactured by the conversion of $CaCO_3$ to CaO by heating (($CaCO_3 \xrightarrow{heat} CaO + CO_2$). The tonnage of cement sold annually provides a good estimate of the CO_2 emitted by this source, not withstanding some impurities in cement. United Nations data show a rapid expansion of global cement manufacture, an increase of 760% since 1950. This growth rate is second only to that of natural gas use, but cement manufacture accounts for only 2–3% of total CO_2 emission. Weather action on exposed limestone also releases CO_2, but estimates of amounts are sketchy.

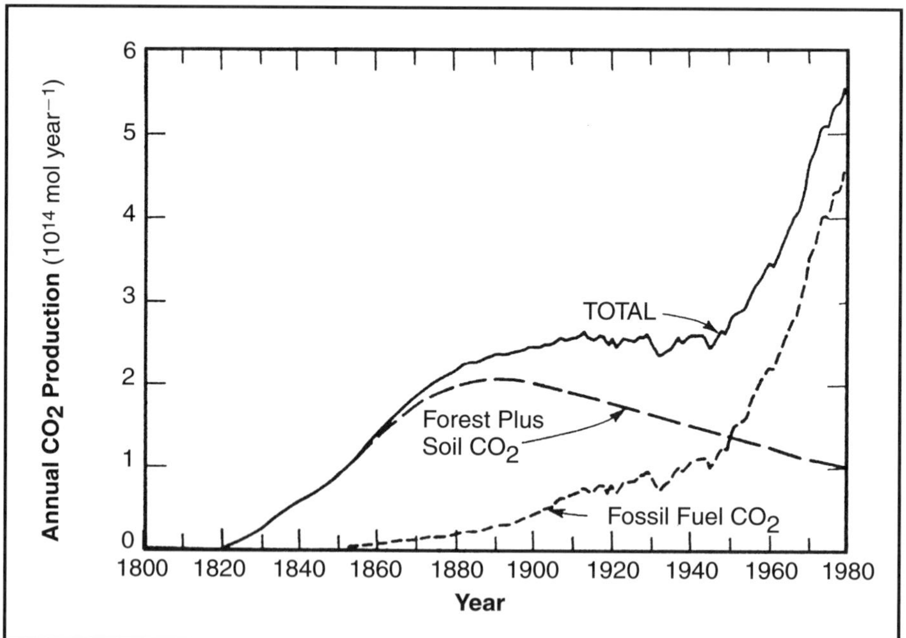

Figure 11.2 ***Annual emission of carbon dioxide to the atmosphere from forests plus soils, and from fossil fuels*** *(Reichle* et al., *1985).*

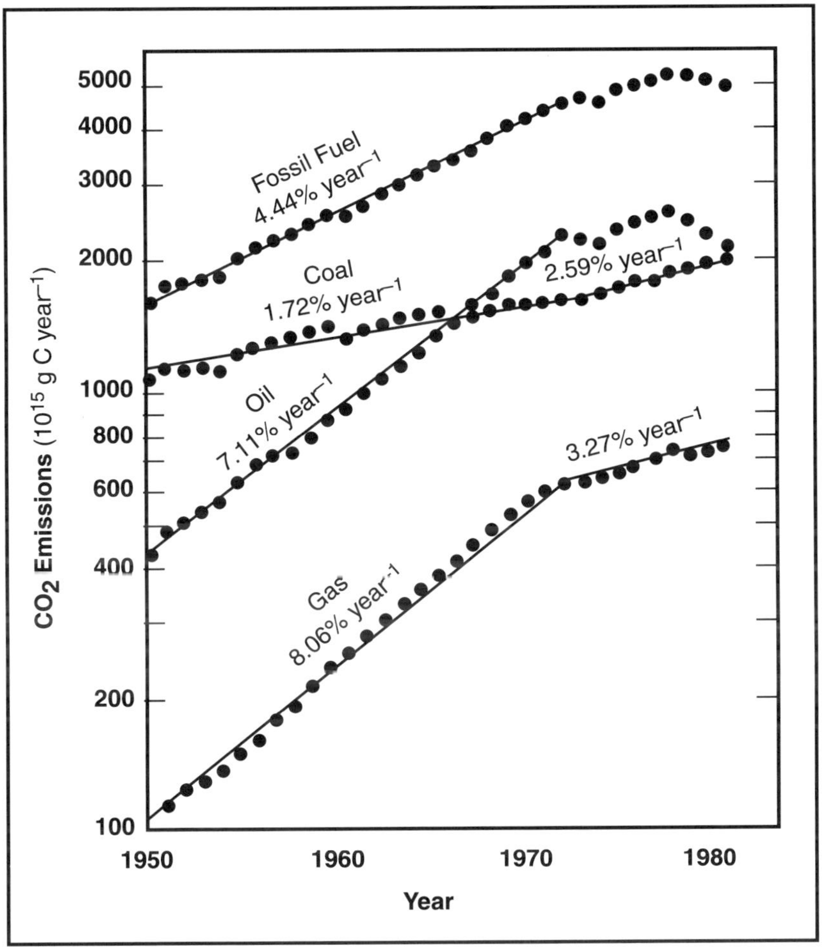

Figure 11.3 ***Annual emissions of carbon dioxide from the use of various fossil fuels*** *(Rotty and Marland, 1986).*

Fossil fuel emission

The atmospheric CO_2 concentration increased to about 295 μmol mol^{-1} by 1900. The increase from Holocene pre-industrial values of about 275 μmol mol^{-1} to that level can be attributed primarily to agricultural activity, especially land clearing. However, the industrial and agricultural revolutions, both currently powered by fossil fuels, caused the 'big increase' in CO_2 emission. Of the estimated 7–8 × 10^9 metric tonnes of carbon currently released each year from all sources, more than 80% can be attributed to the combustion of fossil fuels (Fig. 11.2).

There was a steady exponential increase in CO_2 emissions from fossil fuels up to the 1974 oil crisis (Fig. 11.3). Annual CO_2 emission from oil peaked in 1979 and actually declined in the early 1980s. This decline can probably be attributed to the sharply rising costs of oil, and measures taken to improve efficiency in energy use. In 1986, however, emission from oil again began to rise as cost of oil dropped and the public became less efficiency conscious. Use of natural gas increased about ten-fold faster than other fuels since 1950. CO_2 emissions from liquid and solid forms (used heavily for electricity generation) were approximately equivalent in 1986 (Fig. 11.3).

Overall, the 1990 projection of these emission data may be too low since bigger and more powerful automobiles are back again. The sharp increase in use of natural gas is probably due to domestic heating and industrial usage. This trend is likely to continue because of convenience and competitive pricing.

The role of the ocean in the global carbon cycle

As a major sink and source of CO_2, the oceans are extremely important in the global carbon balance. Most estimates show that 40–50% of net CO_2 emission into the atmosphere is taken up by the oceans. Concentration of CO_2 in sea water is increasing; the carbon concentration in the Atlantic surface water about doubled over the past 10 years. This change in concentration decreases with depth especially after 1 km. Solubility of CO_2 is related to the water temperature and pH. As the oceans warm, they hold less CO_2 in solution, but this effect might be offset by an increase in plant growth rate in surface waters.

Dynamics of Atmospheric CO_2

The best, continuous records of atmospheric CO_2 concentration have been obtained at Mauna Loa Observatory, Hawaii. These records began in 1956 and continue today. Keeling *et al.* (1976) observed that the atmospheric CO_2 concentration varies with season (Fig. 11.1). The seasonal peak (high) occurs during late winter and spring months (in the northern hemisphere), i.e., when the vegetative CO_2 uptake rate is lowest. The seasonal low occurs in the early fall, due to the photosynthetic activity of plants during summer that lowers the atmospheric CO_2 level. Annual variations in atmospheric CO_2 concentration in the stratosphere above about 12 km are small. Other observers have shown that, in the northern hemisphere, the concentration is 8–10 μmol mol^{-1} higher near the ground in spring than the high altitude steady-state concentration. On the other hand, in September, the CO_2 concentration is 10–12 μmol mol^{-1} lower near ground level as compared with high altitude concentrations. Seasonal variation to this extent would hardly be expected in the southern hemisphere, since it consists of 80% ocean compared with 60% ocean in the northern hemisphere.

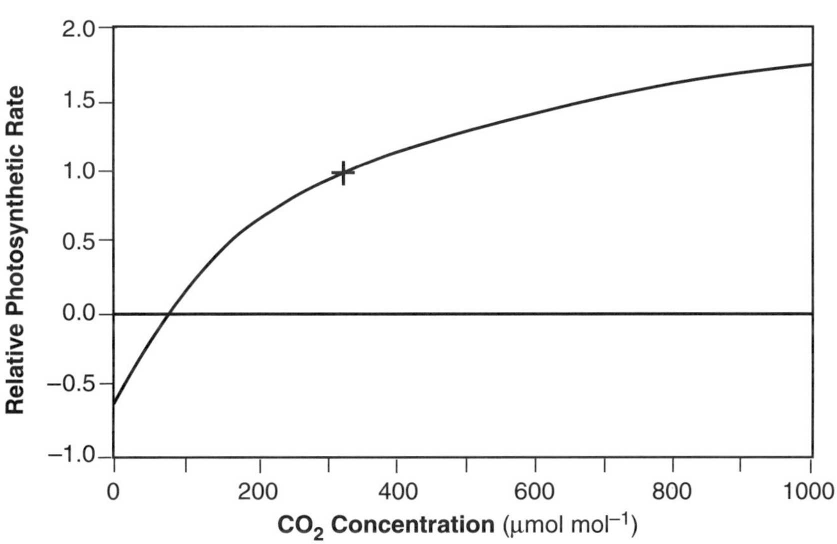

Figure 11.4 ***Whole plant canopy photosynthesis rate at varying atmospheric carbon dioxide concentrations*** *(Allen* et al.*, 1987).*

Since photosynthesis is a strong CO_2 sink and soil plus plant respiration are strong CO_2 sources, diurnal variation in CO_2 concentration near ground level should be expected. This is especially true in plant canopies because they extract CO_2 during the day and release it along with that from soil respiration at night. The highest CO_2 concentration in a maize canopy is in early morning (i.e. predawn). The lowest CO_2 is at noon and early afternoon at the ear level where the greatest concentration of leaf area and highest photosynthetic activity is found. In late afternoon near sunset the atmospheric concentration begins to rise again due to cessation of photosynthetic activity and buildup of respired CO_2.

CO_2 Effects on Plants

Some 31 effects have been attributed to elevated levels of CO_2, ranging from meteorological to various plant physiological processes. From an ecological standpoint the most relevant effect is that of increasing photosynthesis in natural and managed plant production ecosystems.

Photosynthesis

CO_2 is a substrate in the photosynthetic reaction and an increase in CO_2 results in an increase in photosynthesis rate (Fig. 11.4). This is especially true because current ambient CO_2 concentration is below concentrations that saturate photosynthesis. The response to increasing CO_2 levels is hyperbolic rather than linear, and shows a diminishing return with increasing CO_2. Soybean seed yield increased by 32% by doubling the CO_2 concentration from 315 μmol mol^{-1} (ambient) to 630 μmol mol^{-1}.

Photosynthetic and growth differences among species

Three major carbon fixation pathways have been well defined:

1. Ribulose-1,5-bisphosphate carboxylase/oxygenase (RUBISCO) is the CO_2 fixation enzyme in the C_3-type species as represented by soybean, rice, wheat, oat, barley, rye, pulses, and many fruit, vegetable, and woody species.
2. Phosphoenolpyruvate carboxylase enzyme initially fixes CO_2 in the C_4-type species as represented by tropical grasses including maize, millet, sugarcane, and sorghum.
3. Crassulacean acid metabolism (CAM) uses RUBISCO in dark fixation of carbon and is represented by pineapple. The CAM pathway is common among the epiphytes, *e.g.*, certain ferns, Spanish moss, and other 'air plants'.

The most responsive to elevated CO_2 among the three carbon-fixation types has been observed to be C_3 species. In field-grown environments, cotton is especially responsive to CO_2 (Fig. 11.5). The C_3-type species are inefficient at low CO_2 concentrations in large part because O_2 binds on the primary photosynthesis enzyme and competes with CO_2. Furthermore, the C_3 plant has to expend biochemical energy to remove the O_2 from the enzyme and its reaction products in a process called photorespiration. At high concentration, CO_2 competes more effectively with O_2, and thus C_3 plants show a large response to elevated CO_2. The C_4-type species have a built-in CO_2-concentrating mechanism and thus do not experience O_2 competition for CO_2 at the final fixation sites. Therefore, C_4 species respond little to CO_2 concentrations above current atmospheric levels.

The aperture of stomata is governed by the CO_2 concentration of the air, among other factors. Thus, CO_2 can function as a bioregulator of stomata. In general, leaf stomatal conductance for CO_2 and for water vapor is decreased by about 40% by a doubling of CO_2 concentration. However, most studies show that the transpiration rate of whole canopies of plants is decreased only by about 10% by doubled CO_2. When stomata partially close (*e.g.*, 40%), conditions must adjust to achieve a new leaf energy balance. Leaf transpiration rates will tend to decrease because of the greater restriction on water vapor diffusion from the leaf. Consequently, leaf temperature will tend to increase because of less evaporative cooling. However, the higher leaf temperature will, in turn, cause a higher leaf transpiration rate. The end effect of these tightly coupled responses is that the leaf and canopy transpiration rates may be decreased a modest amount by elevated CO_2. However, the decrease in transpiration will not be in direct proportion to the reduction in stomatal conductance.

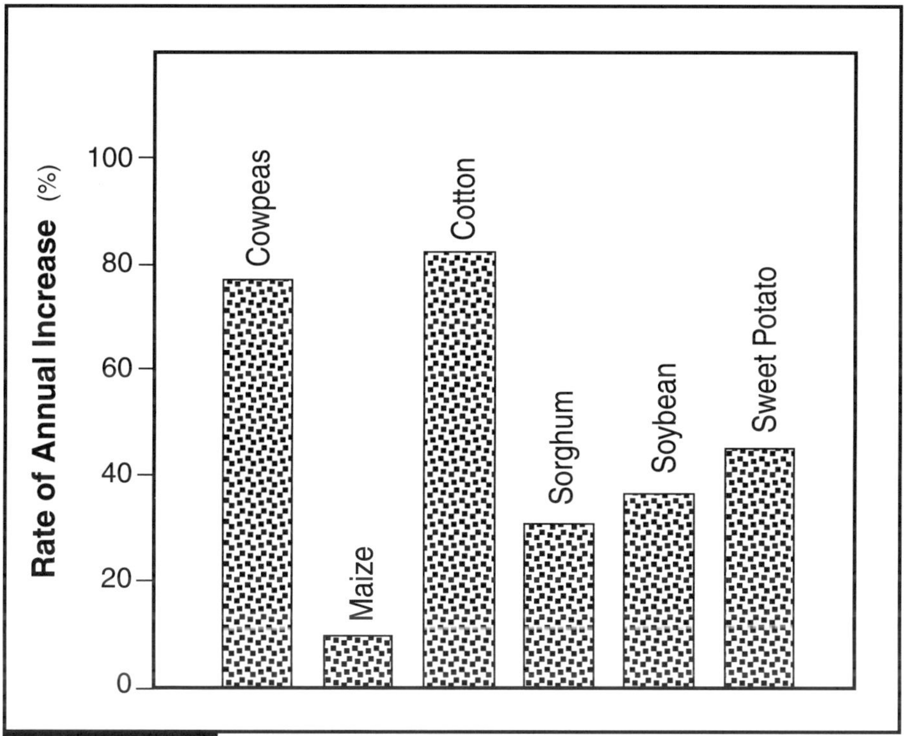

Figure 11.5 ***Yield increase for various crop species as a result of doubling atmospheric carbon dioxide concentration.***

Furthermore, since plants usually grow more leaves or larger leaves when enriched with CO_2, there may be more transpiring leaf surface area.

Water-use efficiency is defined as the ratio of CO_2 uptake to H_2O transpired. Soybean plants grown at 800 μmol mol^{-1} CO_2 had a water-use efficiency that was slightly more than twofold that of soybean grown at 330 μmol mol^{-1} CO_2. However, most of the increase in water-use efficiency was due to an increase of photosynthetic rate of plants grown in elevated CO_2. A much smaller effect on water-use efficiency can be attributed to a reduction in transpiration rate.

Dry matter and seed yields

Total dry weight of soybean and all component parts' dry weights were increased with elevated CO_2 levels. Because there is little change in the ratio of weights among various components, partitioning of photosynthates during plant development appears constant for the species despite CO_2 level. In wheat, increased CO_2 during floral initiation to anthesis resulted in increased kernel number, but a higher CO_2 level increased yield in the grain-filling stage by increasing kernel size. For rice, the number of panicles per plant was the most important factor in increasing rice yield with elevated CO_2. The number of filled grains per panicle and the grain mass per seed were not changed by increased CO_2.

Meteorological Effects of Atmospheric CO_2

There has been a slight but gradual global warming trend since about 1850 in both northern and southern hemispheres (Fig. 10.6). This warming has sometimes been attributed to an increasing greenhouse effect. All general circulation models of climate change predict less warming near the equator than at the poles. The worst probable scenario proposed would result in global warming by as much as 2–10°C. Warming of 10°C would seriously affect climate, ocean levels, glaciers, as well as crop production and food supplies. A temperature increase of 1.5°–5.5°C is frequently projected by doubling CO_2 concentration, but some models predict a temperature increase as low as 0.26°C.

It is important to recognize that there is a broad spectrum of informed opinion on climatic change as related to CO_2 elevation. Diversity of viewpoints appears to emanate from the following facts.

1. Carbon dioxide is only one of several fossil or anthropogenic gases that absorb infrared radiation and produce atmospheric warming. The combined effect of CO_2 and other thermal infrared absorbing gases results in an estimated *effective* 'greenhouse' concentration of 408 μmol mol^{-1}, rather than the reported 351 μmol mol^{-1} for CO_2 alone. Future effects will depend on emissions, or controls, of any one or all of the greenhouse-effect gases.
2. There is a diurnal aspect to 'greenhouse warming'. Current warming trends appear to be due to raising only the minimum (night) temperature, while daytime temperatures may have declined. Mean temperatures have risen less than the projected, i.e., 0.7°C (1.3°F) from models using the current CO_2 level, only a third of the predicted value.
3. Atmospheric aerosols and particulates have increased and they reflect a small amount of the incident solar radiation from the atmosphere. This decrease in solar radiation at the Earth's surface would decrease the global warming potential.
4. There are spatial and temporal warming aspects of global warming. Some continents have been warmer than the global mean for decades, while other continents were cooler for that period. Cities are heat sources and are warmer than the countryside. Currently, population growth is concentrated in cities due to population 'explosion' and 'implosion' (migration from rural areas to cities) and most people live in cities. Does the concentration of people into cities that are largely vegetation-free, non-evaporating surfaces with high emission rates of CO_2 cause spatial or temporal warming of the globe?
5. Oceans mitigate land warming. This view is complicated by the fact that the southern hemisphere has followed the warming trend as projected, while the

change in the northern hemisphere has been lower than projected; the southern hemisphere is only 20% land surface where CO_2 releases are occurring, while the northern hemisphere has 40% land area, so the reverse would have been expected. Also, oceans sequester vast amounts of CO_2 in organic and mineral sediments.

6. While the increasing trend of atmospheric CO_2 concentration is not in dispute, its impact on global warming is. The southern hemisphere appears to have followed closely the projected warming scenario. However, year-to-year temperature variations around the century average temperature in the northern hemisphere are too large to be conclusive. The colder 1970s were followed by the warmer 1980s. One analyst found no change in North America over the past 50 years.

Methane: Rice Paddies, Wetlands, and Other Sources

Sources of methane emissions to the atmosphere

Methane (CH_4) is one of the greenhouse gases, next to CO_2 in importance, that may play a large role in climate change. Recently, methane concentrations have been rising at the rate of 1% per year. The current concentration is about 1.7 μmol mol^{-1}, which is only about 0.48% of atmospheric CO_2 concentration. However, the CH_4 concentration was only about 0.8 μmol mol^{-1} in pre-industrial times. Methane has, therefore, more than doubled in concentration whereas CO_2 has increased by only about 30%.

Methane concentration, like CO_2, is influenced by the biosphere and human industrial activity. Gas bubbles trapped in various layers of Antarctica and Greenland ice over the last 200,000 years have been analyzed from extracted ice cores. These ice core bubbles reveal that both CH_4 and CO_2 concentrations in the atmosphere have followed global temperature changes closely. It is possible that climatic changes may have affected the atmospheric concentrations of these gases, rather than the concentration of the gases governing climate changes.

Scientists have measured rates of release of CH_4 from various sources and have estimated the annual input to the atmosphere. Table 11.1 provides one estimate of the annual efflux to the atmosphere. Paddy rice culture and livestock production (including animal wastes) are two large sources of CH_4 release to the atmosphere.

The greenhouse warming potential of CH_4 is greater than for CO_2. Relative absorption capacity of thermal radiation for a unit mass increase in concentration is 58-fold greater for CH_4 than for CO_2 at the current atmospheric concentrations. The greenhouse warming potential of CH_4 may be less in the long term because of the lower residence time in the atmosphere of CH_4 as compared with CO_2.

Table 11.1. Estimated annual sources and sinks of atmospheric methane (Houghton *et al.*, 1992).

	Tg CH_4 year^{-1}
Sources	
Wetlands	115
Rice paddies	60
Livestock (ruminants)	80
Coal mining, natural gases, and petroleum industry	100
Biomass burning	40
Termites	20
Landfills	30
Animal wastes	25
Ocean	10
Fresh waters	5
CH_4 hydrate	5
Domestic sewage treatment	25
Total	515
Sinks	
Removal by soils	30
Reaction with OH in the atmosphere	460
Total	490
Net input to atmosphere	25

Methane from rice paddies

The scientific assessment of the magnitudes of the various sources and sinks of atmospheric CH_4 continues. Rice paddies have been identified as a primary source of CH_4 emission. Measured rates of CH_4 emissions range from 4.5 to 15.9 g m^{-2} from Texas rice paddies up to as much as 170 g m^{-2} from a Chinese rice field. Aggregated CH_4 emissions from rice paddies could amount to about 85–95 Tg CH_4 per year. This is about 15–20% of the annual flux of CH_4 to the atmosphere. Measurements of CH_4 fluxes from Florida Everglades vegetation showed that emission rates depend on canopy photosynthetic CO_2 fixation rates and showed a pronounced diurnal variation. Cattail has both greater photosynthetic CO_2 uptake rates and CH_4 emission rates than does sawgrass.

Bacterial production of methane is prevalent in most anaerobic environments and it is the result of a specific type of bacterial energy-yielding metabolism. The reaction provides energy by the anaerobic oxidation of inorganic or simple organic compounds. This oxidation is coupled with the reduction of CO_2 to CH_4. Methane is produced only when the land is flooded and soil conditions are oxygen deprived. When soils contain oxygen, CH_4 is not emitted; however, emission returns within a few days after flooding. This is consistent with the strictly anaerobic habit of the organisms producing methane and the requirement of a low reduction–oxidation (redox) potential ($E_h < -200$ mV) for methane production. As an aerated soil becomes waterlogged, oxygen is depleted by respiration and soil microorganisms then use a series of electron acceptors: nitrates, manganous manganese, the ferric–ferrous iron system, sulfate and finally carbon dioxide which is reduced to CH_4.

While advances are being made in our knowledge of the bacteriology and physiology of methanogenesis, the ecology of methanogenic bacteria in the soil is poorly understood. Much more CH_4 may be produced in waterlogged soils than escapes to the atmosphere. Apparently, much of the CH_4 is oxidized by methane-consuming bacteria as it diffuses back along the same pathway as oxygen diffuses to the rhizosphere. Many factors are involved in determining the amount of CH_4 that is ultimately released to the atmosphere, including plant soil parameters and management practices. For example, in rice the pathway for CH_4 emissions from paddies is mainly through specialized air channels in the roots, stems, and leaf sheaths, called aerenchyma. This is the same pathway in the rice plant for oxygen transport to the root system. Natural wetlands plants also have aerenchyma as a part of their anatomy to allow transport of oxygen to submerged organs, and therefore, a transport pathway for the emission of CH_4 to the atmosphere.

Other Industrial Gases

Other industrial gases are also of concern regarding their effects on the environment. Nitrogen gases (NH_3, N_2O, NO, NO_2) released into the atmosphere have increased steadily with the increasing consumption of fossil fuels. Like CO_2, these gases (especially N_2O) also absorb infrared radiation, therefore, they can be classified as greenhouse-effect gases. Human activities are primarily responsible for emissions of N gases, but they are not the sole source. Leaves of well-fertilized crop plants have been shown to gain or lose N as ammonia assimilation or emission. This depends on atmospheric ammonia concentration and physiological status of the plant. Wheat plants have been observed to lose about 6 mg NH_4-N per m^2 of leaf area per day during most of the growing season. The loss rate increased to about 25 mg NH_4-N per m^2 of leaf area per day during crop senescence. Animals emit N gases, especially ammonia.

Heavy industry, especially smelters, emits large quantities of sulfur dioxide (SO_2). A weak acid, H_2SO_3, is formed in moist air, which is seriously eroding buildings and various structures, including such ancient edifices as those of Egypt. Deposition of CO_2 as acid rain can seriously affect terrestrial and freshwater biota and may affect human health. Dry deposition of particulate matter may also occur. Construction of high smoke stacks to reduce local depositions has resulted in distant downwind deposition problems, often in other geopolitical territories. The coniferous forests of northern Finland have been directly damaged from the emissions of two large Russian smelters near the Finnish border. These smelters reportedly released over half a million tons of SO_2 per year.

Like N, sulfur is a constituent of amino acids and is an essential plant nutrient. In the western USA and many areas of the world with little industrial emission, sulfur may be limiting for plant growth. Also, H_2SO_4 is no longer used in the manufacture of phosphatic fertilizers, which exacerbates sulfur deficiency problems in certain areas. Generally, sulfur pollution is restricted to areas with industries that use fossil fuels.

Sulfur, as sulfate particles, can have an effect on the climate. In 1982 the El Chichón volcano (Mexico) spewed some 10×10^7 metric tonnes of aerosol into the stratosphere (20–30 km) of the northern hemisphere. Much of the aerosol was SO_2 and the particles caused a white cloudy haze in the stratosphere. A temperature decrease of 0.2–0.5°C was recorded at the Mauna Loa Observatory in contrast to increased temperature due to greenhouse gases. The Mount Pinatubo eruption in the Philippines in June 1991 also had an effect on global surface temperatures, as well as a reduction in stratospheric ozone.

Additional gaseous pollutants are carbon monoxide, and occasionally, halides, but the greatest concern currently is that of the chlorofluorocarbons (CFCs) of industrial and domestic emissions. The CFCs are used as blowing agents in the manufacture of closed-cell foams and open-cell foams, as refrigerants, aerosol propellants, cleaning or drying fluids, and heat transfer fluids. Sale of aerosols that use CFCs is currently restricted in the USA and in many other countries. Leakage from automobile refrigeration is a major source of CFC pollution in the USA. Many nations have signed an agreement to phase out the use of CFCs (The Montreal Protocol on Substances that Deplete the Ozone Layer, signed 22 September 1987).

The release of CFCs is serious because of their capacity to destroy ozone (O_3) in the upper atmosphere and stratosphere. Ozone provides the protective shield against harmful ultraviolet (UV) radiation, especially the UV-B band (Chapter 8). Disappearance of stratospheric O_3 has become conspicuous over both poles of the Earth, and a general global reduction of stratospheric O_3 is now noted. Excess ultraviolet radiation absorption, especially by light-skinned individuals, is potentially carcinogenic. Recent concern has prompted international scientific and political conferences to be held with the objective of convincing all nations to ban the use of CFCs in aerosols and to work toward other long-range remediations.

Progress has been made toward identifying gaseous products that can be used as a substitute for CFCs. Most of these gases are either CFCs with hydrogen substitution for one or more chlorine atoms (i.e., HCFCs), or fluorocarbons without any chlorine at all (i.e., HFCs). These gases are being evaluated for efficacy, safety for humans, longevity in the atmosphere, effect on stratospheric ozone, and for greenhouse warming potential.

Summary

Since about 1850, the carbon dioxide (CO_2) concentration in the atmosphere has steadily increased at an ever-increasing annual rate (currently about 1.5 $\mu mol\ mol^{-1}\ year^{-1}$). Combustion of fossil fuels (natural gas, oil, and coal) for manufacturing, transportation, and domestic use is the primary source of this upward trend. Agricultural activities and perturbations to natural ecosystems such as logging, forest fires, and land clearing are also contributors to increased levels of CO_2. Other gases (CH_4, N_2O, and CFCs) are increasing and contribute to the global warming potential. Collectively, these gases are called the greenhouse-effect gases since they entrap outgoing radiant energy and warm the planet. They may cause further global warming if their concentrations continue to increase.

Increased atmospheric CO_2 concentration acts in the following important ways: (i) as an aerial gaseous fertilizer, increasing photosynthesis, plant mass, and seed yields; (ii) bioregulation affecting stomata; and (iii) meteorologically, by absorption of infrared radiation emitted from the Earth, causing a rise in surface temperature and certain other climate changes, *i.e.*, the greenhouse effect. Most plant species respond to elevation of CO_2, but C_3 types are more responsive than C_4 types, *e.g.*, cotton is more responsive than maize. For a doubling of atmospheric CO_2, crop yields might increase by 30% for C_3 species and considerably less for C_4 species. Response to CO_2 elevation may be decreased if other growth factors such as temperature are limiting. Water use efficiency is increased by elevated CO_2. Should projected doubling of atmospheric concentration of CO_2 occur, temperature might be expected to increase by about 3°C. Global warming could have major climatic, biological, and social consequences.

Further Reading

Allen, L.H., Jr, Boote, K.J., Jones, J.W., Jones, P.H., Valle, R.R., Acock, B., Rogers, H.H. and Dahlman, R.C. (1987) Response of vegetation to rising carbon dioxide: Photosynthesis, biomass, and seed yield of soybean. *Global Biogeochemical Cycles* 1, 1–14.

Allen, L.H., Kirkham, M.B., Olszyk, D.M. and Whitman, C.E. (1997) *Advances in Carbon Dioxide Effects Research*. ASA Special Publication Number 61. American Society of Agronomy, Madison, Wisconsin.

Baker, J.T. and Allen, L.H., Jr (1993) Contrasting crop species responses to CO_2 and temperature: rice, soybean, and citrus. *Vegetatio* 104/105, 239–260.

Bolin, B. (1977) Changes in land, biota and their importance for the carbon cycle. *Science* 196, 613–615.

Cicerone, R.J. and Oremland, R.S. (1988) Biogeochemical aspects of atmospheric methane. *Global Biogeochemical Cycles* 2, 299–327.

Cure, J.D. and Acock, B. (1986) Crop responses to carbon dioxide doubling: a literature survey. *Agricultural and Forest Meteorology* 35, 127–145.

Gleason, J.F., Bhartia, P.K., Herman, J.R., McPeters, R., Newman, P., Stolarski, R.S., Flynn, L., Labow, G., Larko, D., Seftor, C., Wellemeyer, C., Komkyr, W.D., Miller, A.J. and Planet, W. (1993) Record low global ozone in 1992. *Science* 260, 523–526.

Houghton, J.T., Callendar, B.A. and Varney, S.K. (eds) (1992) *Climate Change 1992: the Supplementary Report to the IPCC Scientific Assessment*. Intergovernmental Panel on Climate Change, Cambridge University Press, Cambridge.

Jouzel, J., Barkov, N.I., Barnola, J.M., Bender, M., Chappellaz, J., Genthon, C., Kotlyakov, V.M., Lipenkov, V., Lorius, C., Petit, J.R., Raynaud, D., Raisbeck, G.,

Ritz, C., Sowers, T., Stievenard, M., Yiou, F. and Yiou, P. (1993) Extending the Vostok ice-core record of palaeoclimate to the penultimate glacial period. *Nature* 364, 407–412.

Kaiser, H.M. and Drennen, T.E. (1993) *Agricultural Dimensions of Global Climate Change*. St Lucie Press, Delray Beach, Florida.

Keeling, C.D., Bacastow, R.B., Bainbridge, A.E., Ekdahl, C.A. Jr, Guenther, P.R., Waterman, L.S. and Chin, J.F.S. (1976) Atmospheric carbon dioxide variations at Mauna Loa Observatory, Hawaii. *Tellus* 28, 538–551.

Miller, S. (1988) Agriculture and the greenhouse effect. *Agricultural Research* 36, 7–9.

Post, W.M., Peng, T.-H., Emanuel, W.R., King, A.W., Dale, V.H. and DeAngelis, D.L. (1990) The global carbon cycle. *American Scientist* 78, 310–326.

Ramade, F. (1984) *Ecology of Natural Resources*. Pitman Press, Bath.

Reichle, D.E., Trabalka, J.R. and Solomon, A.M. (1985) Approaches to studying the global carbon cycle. In: Trabalka, J.R. (ed.), *Atmospheric Carbon Dioxide and the Global Carbon Cycle*. DOE/ER-0239. US Department of Energy, Carbon Dioxide Research Divison, Washington, DC, pp. 15–24.

Rogers, H.H., Thomas, J.F. and Bingham, G.E. (1983) Response of agronomic and forest species to elevated atmospheric carbon dioxide. *Science* 220 428–429.

Rotty, R.M. and Marland, G. (1986) Fossil fuel combustion: recent amounts, patterns, and trends of CO_2. In: Trabalka, J.R. and Reichle, D.E. (eds), *The Changing Carbon Cycle*. Springer-Verlag, New York, pp. 474–507.

Rozema, J., Lambers, H., Van de Geijh, S.C. and Cambridge, M.L. (eds) (1993) *CO_2 and Biosphere*. Kluwer Academic Publishers, Dordrecht, The Netherlands.

Stolarski, R.S. (1988) The Antarctic ozone hole. *Scientific American* 258, 30–36.

Trabalka, J.R. (1985) *Atmospheric Carbon Dioxide and the Global Carbon Cycle*. Office of Energy Research, US Department of Energy, DOE/ER-0239.

Woodwell, G.M. and Mackenzie, R.T. (1995) *Biotic Feedbacks in the Global Climatic System: Will the Warming Feed the Warming?* Oxford University Press, New York.

Appendix of Scientific Names

Alfalfa	*Medicago sativa*
Almond	*Prunus amygdalas*
Apple	*Pyrus malus*
Apricot	*Prunus armeniaca*
Bahia grass	*Paspalum notatum*
Banana	*Musa sapientum*
Barley	*Hordeum vulgare*
Bermuda grass	*Cynodon dactylon*
Big bluestem grass	*Andropogon gerardi*
Birdsfoot trefoil	*Lotus corniculatus*
Black medic	*Medicago lupulina*
Broadbean	*Vicia faba*
Bromegrass	*Bromus inermis*
Buckwheat	*Fagopyrum vulgare*
Cabbage	*Brassica oleracea*
Cacao	*Theobroma cacoa*
Cassava	*Manihot esculenta*
Cattail	*Typha* sp.
Celery	*Apium graveolens dulce*
Cherry	*Prunus* spp.
Chickpea	*Cicer arietinum*
Cocklebur	*Xanthium pennsylvanium*
Coffee	*Coffea arabica*
Common bean	*Phaseolus vulgaris*
Cotton	*Gossypium hirsutum*
Cowpea	*Vigna unguiculata*
Dandelion	*Taraxacum officinale*
Eggplant	*Solanum melongena*
Fir	*Abies* spp.
Flax	*Linum usitatissimum*
Indiangrass	*Sorghastrum nutans*

Kentucky bluegrass	*Poa pratensis*
Kudzu	*Pueraria lobata*
Lettuce	*Lactuca sativa*
Little bluestem grass	*Andropogon scoparius*
Lentil	*Lens esculenta*
Maize	*Zea mays*
Mango	*Mangifera indica*
Millet	*Pennisetum glaucum*
Napier grass	*Pennisetum purpureum*
Oak	*Quercus* spp.
Oat	*Avena sativa*
Okra	*Abelmoschus esculentus*
Orange	*Citrus sinensis*
Orchard grass	*Dactylis glomerata*
Pea	*Pisum sativum*
Peach	*Prunus persica*
Peanut	*Arachis hypogaea*
Pear	*Pyrus communis*
Pecan	*Carya illinoensis*
Pigeon pea	*Cajanus cajan*
Pineapple	*Ananas comosus*
Potato	*Solanum tuberosum*
Radish	*Raphanus sativa*
Rapeseed	*Brassica napus*
Redwood	*Sequoia sempervirens*
Rice	*Oryza sativa*
Rubber	*Hevea brasiliensis*
Rye	*Secale cereale*
Saw grass	*Cladium* spp.
Sorghum	*Sorghum bicolor*
Soybean	*Glycine max*
Strawberry	*Fragaria* spp.
Sugarbeet	*Beta vulgaris*
Sugarcane	*Saccharum officinarum*
Sweet potato	*Ipomoea batatus*
Sweet clover	*Melilotus indica*
Switchgrass	*Panicum virgatum*
Tea	*Thea sinensis*
Teosinte	*Zea mexicana*
Timothy	*Phleum pratense*
Tobacco	*Nicotiana tabacum*
Tomato	*Lycopersicum esculentum*
Triple awngrass	*Astrida oligantha*
Triticale	*Triticale* spp.
Wheat	*Triticum aestivum*
White clover	*Trifolium repens*
Yam	*Dioscorea* sp.

Index

Africa
 food requirements 4
 malnutrition 9
 population 2
Agricultural ecosystems 51–55
 industrialized agriculture 54–60
 livestock farming 53–54
 sedentary subsistence farming 52–53
 shifting cultivation 51–52
 soil 80–81
Atmosphere 169–183
 circulation 156–162
 industrial gases 181–182
 radiation transmissivity 124–126
Available soil water 111–112

Biotechnology 45–46
Birth rates 2–3
Borlaug, N. 10, 72

C_3 and C_4 species 67
Carbon dioxide 169–179
 atmosphere dynamics 174–175
 fossil fuels 173–174
 global cycle 171–174
 meteorological effects 178–179
 photosynthesis 107–108, 175–177
Cement manufacture 172
China
 food requirements 4, 7–8
 population 2
 water 14
Chlorophyll 66
Climate 15–16, 162–166
 change 165–166
 classification 164–165
Cotton 5–6
Crops
 centers of origin 37–40

Crops *continued*
 chilling and cold tolerance 150–151
 collection 40–42
 development 144–147, 149–151
 domestication 33–35
 drought response 112–116
 genetics 42–46
 international research centers 41
 nutrients 55–57
 origins 31–33
 photosynthesis 176
 spread 32–33
 yields 9–11, 55–57, 69–77
  simulation 75

Domestication of crops 33–35
Diversity
 biodiversity 11–12
 genetic 37–42
 plant diversity 33, 36
Drought 112–116
 avoidance 115–116
 tolerance 114–115

Ecology 19–30
 food chains 24–25
 nutrient cycling 26–27
 primary production 25–26
 succession 23–24
Ecosystems 19–29
 agricultural 21–22
 comparisons 27–29
 perturbation 23
 production 75–77
 structure 21–22
 succession 23–24
 sustainability 28–29
Energy balance 137–138
 latent heat 138

Energy balance *continued*
radiational 131–133
sensible 137–138
Erosion 13, 99–101
Evaporation 105–107

Fertilizer 57, 93–98
Fiber requirements 5–7
Food chains 24–25
Food gatherers 49–50
Food production 4–5
environmental consequences 11–16
land losses 12–14
water 14–15
Food requirements 4
economic considerations 7–9
Forests
carbon dioxide 172
clearing 52
production 6–7
rainforests 20–21
Fossil fuels 121–122, 173–174

Genetics
biotechnology 45–46
diversity 37–42
founder principle 44
mutations 42–43
ploidy 44–45
recombination 43
selection 43–44
transformation 46
uniformity 36–37
Green revolution 9–10, 72, 130
Greenhouse effect 16, 133, 169–171

Harvest index 72–73

Industrialized agriculture 58–60
International crop research centers 41
Intertropic convergence zone 157–158, 162–165
Irrigation 116–119

Land losses 12–14
Landraces 36
Leaf area index 64–65
Liebig, J. 54–55

Maize 36, 38–39, 71, 114
Mayan civilization 36
Membranes 105–106
Methane 179–181
Monsoon 159–161
Mulch 140–142

Nitrogen 74
Nutrient
crop yield 55–57
cycling 26–27
mobility in soil 89–91
overfertilization 98
soil 88–91
transport in soil 89

Ozone hole 130, 158, 182

Photoperiodism 128–130
Photosynthesis 66–68, 147–148, 175–176
Photosynthetically active radiation 64, 122–123
Plant collection 40–42
Polyploidy 44–45
Population 1–4
birth rate 2–3
doubling time 2
early increases 1–2
malnutrition 9
modern increases 1–2
Potato 36–37, 53
Precipitation 162

Radiant energy 121–133
energy balance 131–133
solar radiation 63–65, 122–124
thermal infrared 131–132, 169–171
ultraviolet radiation 130–131
Radiation use efficiency 66–68
Radiational frost 132–133
Rainforests 20–21
Ribulose-1,5-bisphosphate carboxylase 66
Rice 55–56, 59–60, 76–77
methane 180
Roots
growth constraints 88–89
soil chemistry 92–93
soil compaction 91
sorption zone 90
surface sorption zone 90–91

Salinization 118
Soil 79–101
cation exchange capacity 84–85
description 81–88
erosion 13, 99–101
fertilization 93–98
horizons 83–84
nutrient 88–91
overfertilization 98
pH 92–93
root growth constraints
soil chemistry 92–93
soil compaction 91
runoff 99

Soil *continued*
  structure 87–88
  taxonomy 82–83
  temperature 138–142
  testing 93–98
  texture 86–87
  water 85–86, 99, 111–112
Solar radiation 15
  albedo 156
  Earth's surface 126–128
  interception 63–65
Soybean 129
Stomata 107–110, 113, 115, 176
Stress
  nitrogen 74
  temperature 74
  water 74

Temperature 74, 135–152
  air 142–144
  cardinal temperatures 135, 146
  chilling and cold tolerance 151
  global 135–136
  hardiness 150–151
  photosynthesis 70, 147–148
  radiational frost 132–133
  soil 138–142
Thermal units 146–147
Transpiration 107–110, 176

Ultraviolet radiation 130–131

Vavilov, V.I. 39–40
Vernalization 145–146

Water 14–15, 74, 103–119
  available soil water 111–112
  erosion 99
  evaporation 105–107
  physical structure 103–104
  precipitation 162
  quality 98
  soil 85–86, 111–112
  solvent 104–105
  transpiration 107–110
  transport in plants 110
Weather 15–16, 155–162
Wheat 38, 54, 72–73, 149
Wind
  Chinook wind 161
  Coriolis effect 156
  erosion 99–100
  Foehn winds 161
  global surface 157–158
  Intertropic convergence zone 157–158, 162–165
  monsoon 159–161

STARTING UP YOUR OWN BUSINESS